Build and Code Creative Robots with LEGO BOOST

Unleash your creativity and imagination by building exciting robotics projects

Ashwin Shah

BIRMINGHAM—MUMBAI

Build and Code Creative Robots with LEGO BOOST

Group Product Manager: Wilson D'souza

Publishing Product Manager: Mohd Riyan Khan

Senior Editor: Arun Nadar

Content Development Editor: Sulagna Mohanty

Technical Editor: Arjun Varma

Copy Editor: Safis Editing

Project Coordinator: Shagun Saini

Proofreader: Safis Editing

Indexer: Pratik Shirodkar

Production Designer: Nilesh Mohite

First published: November 2021

Production reference: 1131021

Published by Packt Publishing Ltd.
Livery Place
35 Livery Street
Birmingham
B3 2PB, UK.

ISBN 978-1-80107-557-2

www.packt.com

Foreword

I am excited to write the foreword for this book, *Build and Code Creative Robots with LEGO BOOST*, which is specifically designed for kids aged 8 to 12 years as well as for LEGO enthusiasts. Today, robotics is the key to executing the industrial revolution 4.0 across the globe. I see robotics as an interdisciplinary branch that includes a lot of learning and application from control systems, coding, and mechanics! If students learn about robotics during their time at school, they can get exposure to their field of choice in a fun and hands-on way, which can shape their career in a much better way. In this book, Ashwin has tried to build a strong base for robotics learning using the LEGO BOOST kit, with increasing levels of complexity. If you follow these chapters one after the other, I am sure that you will be able to build and code your own robots with unique mechanisms. The use of the Scratch programming language to code your BOOST robot is a welcome move from Ashwin and it will help you all to easily code your robots since many of you will be using Scratch programming at school. I am sure you will be able to complete the capstone project and that it will excite you to explore the challenging yet amazing world of various robotics competitions for your age group.

You will gain a lot of technical knowledge to build and code robots and will be able to innovate further using this robotics kit. Remember one thing: the basics that you will learn throughout this book will help you build on your robotics knowledge when you gradually explore more challenging robotics kits, sensors, and programming languages in the times to come.

Wishing you all a very happy and fun-filled time while building and coding these robots.

Mr. Manoj Patel

Director and CEO, RoboFun Lab Pvt. Ltd.

Contributors

About the author

Ashwin Shah is an electronics and communication engineer from India. Teaching is his passion and he started working at the tender age of 16 as a doubt solver for younger kids at after-school classes. Today, he is a successful edupreneur who runs his own STEM-based robotics, coding, and IoT institute – RoboFun Lab. Being one of the pioneers of STEM education in India, he has taught over 3,000 students and trained 200+ educators thus far. He has trained 100+ teams for prestigious STEM competitions with 50+ national and 12+ international awards in the bag. Most of his students are now pursuing careers/ education in STEM at prominent universities. Ashwin was a state-level badminton player during his college days.

About the reviewer

Anita Kumari is a STEM and robotics educator with 13+ years' experience in schooling students. Her major interests are LEGO robotics, STEM, coding, and Arduino. Born and raised in the beautiful Kullu valley of Himachal Pradesh, she studied electronics and communication engineering at the Govt. Polytechnic College, Kandaghat, HP. Working in both the trainer and STEM content development roles has given her broad experience of writing content for school children that is engaging and exciting at the same time.

I'd like to thank my parents, siblings, relatives, friends, and mentors for guiding and supporting me. I greatly appreciate my in-laws, thank you for always being supportive and adjusting to my odd routines.

To my daughter, Sunanda: Even at the age of 8 years, you never trouble me while I am working. I'd also like to thank Packt Publishing for the opportunity to review this wonderful book.

Table of Contents

7

Building a Helicopter

8

Building R2-D2

9

Building an Automatic Entrance Door

10

Building a Candy Dispenser Robot

11
Building a Color-Sorter Conveyor Belt

12
Building a BOOST Racing Car

13
Final Challenge

Bonus Chapters

14

The Grabbing Robot

15

Obstacle Avoidance Robot

16

The BOOST Humanoid

17

The Moon Rover

Other Books You May Enjoy

Index

Preface

Build and Code Creative Robots with LEGO BOOST teaches a range of interesting robotics projects with detailed building instructions. You'll learn how to use motors, sensors, and Scratch programming to build modern-day smart robots while developing your STEM skills in a fun way.

Who this book is for

This book will help 7-12-year-old children who want to learn robotics with LEGO BOOST to develop their creativity, logical thinking, and problem-solving skills. Teachers, trainers, and parents who wish to teach robotics with LEGO BOOST and Scratch will also find this book useful.

What this book covers

Chapter 1, Introduction to the LEGO BOOST Kit, introduces you to the fascinating world of LEGO BOOST and the infinite possibilities of creation that it comes with. You shall understand how the hub works and how to connect the hub to your tablet.

Chapter 2, Build Your First BOOST Robot – Tabletop Fan, will show you how to build a simple table fan using the LEGO bricks and BOOST Hub given in the kit. You will be introduced to the world of programming and given a basic task to turn the fan on and off. A challenge to change speeds is provided in the *Pursue* section.

Chapter 3, Moving Forward/Backward Without Wheels, covers one of the most interesting ways to learn about the basics of movement and the importance of wheels – moving without wheels! With open-ended discussions, you shall explore various ways in which an object can move even in the absence of wheels. You will tinker with some cool activities to control the robot's movement in the absence of wheels!

Chapter 4, LEGO BOOST Rover, covers how to build your first robot with wheels. More concepts of programming, such as various kinds of turns, will be introduced in this chapter. You will be given more programming tasks, such as making robots move in specific shape patterns and eventually using loops to reduce data redundancy.

Chapter 5, Getting into Gear – My First Geared Robot, introduces you to the various kinds of gears available in the BOOST kit with their specific usage. You shall understand the concepts of gear ratio and gearing up and gearing down in detail and relate them to the term "torque." You will then build your own geared car with various gear combinations and explore firsthand how different gear combinations affect movement.

Chapter 6, Building a Forklift, covers how to build your own forklift robot and understand how a load is moved easily in industry using this vehicle. Your forklift will be completely autonomous and will pick up a load from one place, travel to the destination, and place it.

Chapter 7, Building a Helicopter, covers building a helicopter. We have probably all seen a helicopter and a few of us would have taken a ride in one too. How about building your own helicopter with rotating blades that can move from one place to another? You will be taught the basic concepts of flying and activities will be set accordingly.

Chapter 8, Building R2-D2, covers building an R2-D2 robot. Those of us who are Star Wars fans know R2-D2. With this activity, you will apply your knowledge of gears to create the movement of R2-D2 and complete some fun challenges.

Chapter 9, Building an Automatic Entrance Door, provides yet another simple but effective application of a distance sensor – automated entrance doors. You will build your own entrance door with the help of a pulley mechanism as well as a color sensor in this project.

Chapter 10, Building a Candy Dispenser Robot, teaches you how a color sensor works in real life. You shall build a cool candy dispenser robot that will drop different-colored candies based on the color of the LEGO brick detected by the color sensor.

Chapter 11, Building a Color Sorter Conveyor Belt, teaches you how to build a cool industrial application, a conveyor belt with a basic robotic arm. It will be able to sort three different colors at three different places using a color sensor.

Chapter 12, Building a BOOST Racing Car, covers how to build your own race car with a steering wheel mechanism. You shall learn about advanced concepts in programming and use the steering option to control this car. You will make this an autonomous car capable of traversing a complex path.

Chapter 13, Final Challenge, provides a final challenge that you will have to use your own creativity, logical thinking, and problem-solving skills to crack. You will have to build and code your own robot with no guidance from the book except hints and some common references. This will help students and parents understand how your learning has progressed through this book.

Chapter 14, The Grabbing Robot, covers how to build your own grabber robot that can grab things from one point and place it elsewhere. Like the forklift, this robot will be capable of holding objects of any shape. Students will learn about the complex application of gears and build a sturdy robot capable of grabbing and displacing.

Chapter 15, Obstacle Avoidance Robot, introduces you to the world of sensors and explains how color sensors work along with various practical examples. You will be introduced to the programming logic when sensors are involved.

Chapter 16, The BOOST Humanoid, covers how to build a humanoid structure using your BOOST kit and attach a color sensor to it at the bottom. You will learn about the basic concept of line following and then program the robot to follow the black line path given.

Chapter 17, The Moon Rover, covers how to build a moon rover that will be capable of sending sensed data to the main station and perform tasks such as dropping water in barren land and collecting samples from green land.

To get the most out of this book

To effectively perform all the projects hands-on, you must have a LEGO BOOST kit along with a laptop for programming. Basic knowledge of the Scratch programming language would be an added advantage.

Software/hardware covered in the book	OS requirements
LEGO Boost kit External motors and sensors	macOS, Windows, Chrome OS, or Android
Scratch 3.0 programming language	

If you are going to use Scratch 3.0 online, you must create an account there so that you can effectively save your programs and use them for future reference. If you are going to install it on your computer, please download it from here: `https://scratch.mit.edu/download`. You also need a set of six AAA-sized rechargeable batteries for your BOOST Hub.

After you are done reading this book, I would love to see you building more complex robots using your own creativity and innovation. Building a 3D printer is a worthwhile option to consider. You can also explore various LEGO-based robotics competitions and aim to participate in them to showcase your creativity and innovation.

Download the color images

We also provide a PDF file that has color images of the screenshots/diagrams used in this book. You can download it here: `https://static.packt-cdn.com/downloads/9781801075572_ColorImages.pdf`.

Conventions used

There are a number of text conventions used throughout this book.

`Code in text`: Indicates code words in text, database table names, folder names, filenames, file extensions, pathnames, dummy URLs, user input, and Twitter handles. Here is an example: "Simply swap the motor ports in the `if else` condition and the robot will follow the left edge of the line."

A block of code is set as follows:

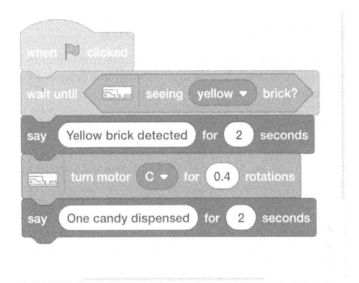

Figure 10.61

Bold: Indicates a new term, an important word, or words that you see onscreen. For example, words in menus or dialog boxes appear in the text like this. Here is an example: "Display **Green brick detected** before dispensing. Once the candies are dispensed, display **Three candies dispensed**."

> **Tips or important notes**
> Appear like this.

Get in touch

Feedback from our readers is always welcome.

General feedback: If you have questions about any aspect of this book, mention the book title in the subject of your message and email us at customercare@packtpub.com.

Errata: Although we have taken every care to ensure the accuracy of our content, mistakes do happen. If you have found a mistake in this book, we would be grateful if you would report this to us. Please visit www.packtpub.com/support/errata, selecting your book, clicking on the Errata Submission Form link, and entering the details.

Piracy: If you come across any illegal copies of our works in any form on the Internet, we would be grateful if you would provide us with the location address or website name. Please contact us at copyright@packt.com with a link to the material.

If you are interested in becoming an author: If there is a topic that you have expertise in and you are interested in either writing or contributing to a book, please visit authors.packtpub.com.

Reviews

Please leave a review. Once you have read and used this book, why not leave a review on the site that you purchased it from? Potential readers can then see and use your unbiased opinion to make purchase decisions, we at Packt can understand what you think about our products, and our authors can see your feedback on their book. Thank you!

For more information about Packt, please visit packt.com.

Share Your Thoughts

Once you've read Build and Code Creative Robots with LEGO BOOST, we'd love to hear your thoughts! Scan the QR code below to go straight to the Amazon review page for this book and share your feedback.

https://packt.link/r/1801075573

Your review is important to us and the tech community and will help us make sure we're delivering excellent quality content.

1
Introduction to the LEGO BOOST Kit

Welcome to the first chapter of this book! I am sure you all are excited to learn about your BOOST kit and start building new robots with the kit in each new chapter! Just before we begin with the construction of our first robot, let's try to learn the basics!

In this chapter, you shall be unboxing your BOOST kit and exploring various electronic and non-electronic parts given in the kit in a fun and hands-on way. We will cover the following topics in this chapter:

- The difference between machines and robots
- Using the various electronic and non-electronic parts that come with your BOOST kit
- Building your first model with this BOOST kit

Technical requirements

In this chapter, you will need the following:

- A LEGO BOOST kit with six AAA batteries, fully charged

Wonders with LEGO BOOST

The LEGO BOOST kit comes heavily loaded with electronic and non-electronic parts. If you learn how to use this kit properly, the sky is the limit. The BOOST kit comes with three motors, built-in gyro sensors, and an external (a separate electronic piece that can be attached to the BOOST Hub) color sensor as well as an ultrasonic sensor. You can make some cool creations with this kit, such as the following:

- A 3D printer
- A humanoid
- A color sorter
- A line follower

You can stretch your creativity to its limit using this BOOST kit to build and code anything and everything that you can imagine.

The difference between machines and robots

You might have come across the question "*How do you differentiate between a machine and a robot?*" Let's try to understand this with a simple example of a fan and an air conditioner.

Imagine how a fan works. When you turn on the switch, the fan starts working, and when you turn off the switch, the fan stops working. In a simple statement, a fan takes an input (turn on/off the switch) and directly gives us an output (it either starts moving in a clockwise direction in the on condition or stops working when the switch is off).

A machine is something that takes an input and directly gives you an output. Can you think of at least four such machines in and around you and write them down?

Names of the machines around you:

1. _____
2. _____
3. _____
4. _____
5. _____

The following figure represents the input and output mechanism in a fan:

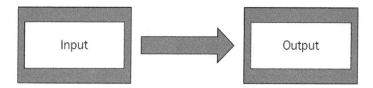

Figure 1.1 – Input/output mechanism of a fan

Now, let's try to understand how an air conditioner works:

1. You turn on the main switch and turn on the air conditioner.

2. You set the desired temperature for the air conditioner to maintain – say, for example, 26 degrees Celsius.

3. If your room temperature reaches more than 26 degrees Celsius, your air conditioner will throw cool air in the room. If the room temperature is less than 26 degrees Celsius, your air conditioner will throw normal air in the room.

So, how is it different from a fan? In your air conditioner, you have a temperature sensor that measures the temperature of the room all the time and a processor that acts based on the input received from this temperature sensor. So, what is happening here is the following:

1. **Input**: You set up the temperature through your air conditioner remote.

2. **Process**: The air conditioner's processor processes this input and compares it with the input received from the room temperature sensor and decides the action accordingly.

3. **Output**: The processor commands the air conditioner to either throw cool air (if the room temperature is more than the set temperature) or normal air (if the room temperature is less than the set temperature).

Can you think of four such examples that can be defined as robots in and around you except an air conditioner?

Names of the robots around you:

1. _____

2. _____

3. _____

4. _____

5. _____

The following figure shows the mechanism of an air conditioner:

Figure 1.2 –Working mechanism of an air conditioner

Similarly, you will be applying the same principles with your LEGO BOOST kit, where you will be building a robot to solve some specific problems. Just before we unbox our BOOST kit, I wanted to highlight some facts about everyone's favorite LEGO sets (source – https://www.goodhousekeeping.com/life/parenting/g2775/facts-about-legos/):

- LEGO is the world's largest manufacturer of wheels.

- Every single LEGO brick made since 1958 can still be joined together.

- Every person on earth owns an average of 86 LEGO bricks.

- During the holiday season, 28 LEGO sets are sold every second across the globe.

- If LEGO mini figures were real people, they would form the highest population in the world.

Introduction to the electronic and non-electronic parts of the BOOST kit

Let's first understand the various electronic parts given in the LEGO BOOST kit with specific functionality:

1. **BOOST Hub**: This hub is the brain of your robot (the microprocessor of the hub serves as the brain). The hub is connected to your tablet via Bluetooth. The hub has two built-in motors (motor A and motor B) as well as an built-in tilt (gyro) sensor. It needs six AAA-sized batteries to operate. It also has two input/output ports:

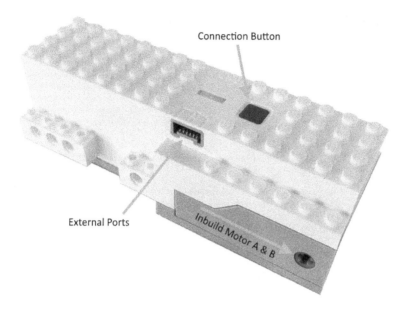

Figure 1.3 – The BOOST Hub

2. **Interactive motor**: The interactive motor comes with an built-in rotation sensor. This allows us to check the number of rotations/the speed at which the motor is moving. This feature also allows us to add precision to the robot's programming; for example, you can move forward or backward for a specific distance/rotation, allowing the same stopping position of the robot each time:

Figure 1.4 – Interactive motor

3. **Color and distance sensor**: The BOOST kit comes with a two-in-one sensor that can act as a color as well as a distance sensor:

Figure 1.5 – Color/distance sensor

4. **Bluetooth Low-Energy (BLE) module**: This module comes fitted in the hub itself. You will be able to connect your BOOST Hub to tablets and Android phones if your device has Bluetooth version 4.1 or higher. If you have an older device, it is recommended to use a BLED112 Bluetooth dongle to connect the hub to your tablet.

Apart from these electronics, your BOOST kit comes with 873 non-electronic LEGO parts. So, these 3 motors, 3 sensors, and 873 non-electronic LEGO parts give you immense flexibility to build anything and everything that you can imagine. In this book, you will be building 18 unique robots with increasing complexity levels.

> **Important note**
>
> You can always refer to `https://www.bricklink.com/catalogItemInv.asp?S=17101-1` for the inventory of your BOOST kit with the original name of each part of the kit.

Now that you have had an overview, I have listed 10 basic parts from this kit in the following table with their names and an example application. Can you play around with these parts and try to write one more application for each of them (under the **Functionality #2** heading)? This will allow you to gain more knowledge of the parts and how to use them in future projects.

Just for your information, one module in LEGO is equal to one hole on a beam. So, if you have a straight beam with five holes in it, it is called a 5M beam, where M stands for module. Similarly, you can measure the size of an axle by placing it against the longest beam. The number of holes the axle covers is the size of the axle. For instance, if the length of an axle is equal to five holes, it is called a 5M axle (where M stands for module):

Object Image	Actual Name of the Object	Functionality #1	Functionality #2
	3 X 5 L connector	Used to connect a motor to the hub.	Used to connect two pieces at 90 degrees.
	Element separator	This brick separator is used to separate two or more studded bricks without much effort.	
	4 X 6 brick	This piece is used in many ways – right from creating a sturdy chassis to connecting multiple pieces across this rectangular frame for some specific structure.	
	36-tooth double conical wheel (bevel gear)	Gears are used to change the direction and magnitude of the applied force. It can be used in chassis, grabbing, and other similar mechanisms.	

Object Image	Actual Name of the Object	Functionality #1	Functionality #2
	T-beam 3X3	As the name suggests, this piece is used to give strong connections and support in a T shape.	
	Wheel with grip	Wheels are used to make the movement of any object easy! For example, in the chassis of a car/robot to make it move easily from one place to another.	
	Axle	A wheel is always connected with an axle for accurate movement. Wheels and axles together make one of the six simple machines.	
	3M friction peg	This piece is used to connect three beams/LEGO pieces at the same time and in the same spot.	
	2M friction peg	This piece is used to connect two LEGO pieces at the same time and in the same spot. Since this is with friction, it provides sturdy connection.	
	2M frictionless peg	This piece is used to make loose, frictionless connections between two LEGO elements.	

There are many more pieces in your kit that we shall explore further in the upcoming projects and understand their uses.

The importance and efficient usage of various pegs

To build your robot, you will have to connect different pieces to each other. As we use nuts and bolts in real life to connect two things, we will be using pegs in the BOOST kit for connections.

> **Tip**
>
> Always use at least two pegs to make any connection sturdy! Do you know that the two module pegs are of two different types? The one in black is a friction peg and the one in gray is a frictionless peg.

As you can see in the following photo, both the LEGO pieces connected with a single peg are not sturdy. They can move easily. This is not the kind of connection we would like to have in our robot, right?

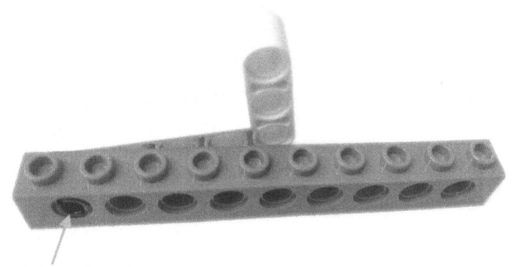

Single-Peg Connection

Figure 1.6 – Single-peg connection

In the following photo, we can see that both the LEGO pieces are now connected with two pegs! The structure is sturdy – the kind of connection that we shall need in our robot:

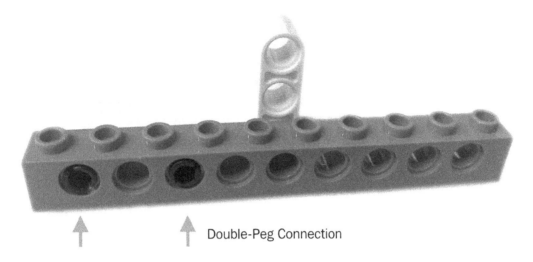

Figure 1.7 – Double-peg connection

Now that you know the part names and the usage of most of them, let's move on to the challenge section, where you will have a go at a free-play activity.

Time for a challenge

Now that you know the names and applications of these LEGO parts, let's try to use them in an effective way with a free-play activity.

Activity #1

Build the tallest and sturdiest tower possible using all the LEGO bricks you have in this set!

All tall towers are named something or the other, such as Burj Khalifa in Dubai and Willis Tower in the US. Since you have also built a tall tower, try to give it a name of your choice! The name you have chosen for your tower is _____.

Now, try to add wheels and make this tower move!

Figure 1.8 – Sample tower built using the LEGO bricks available in the BOOST kit

Did your tower fall? If yes, make it sturdier to ensure that it stays stable even while moving!

Summary

In this chapter, you have understood the difference between machines and robots. You also got to unbox your LEGO BOOST kit and understood in detail all the electronic and non-electronic parts that come with your kit. You also explored the LEGO parts and understood the application of some of the key parts. You should be able to use your LEGO parts effectively and build simple basic structures out of them, as well as identifying the electronic parts correctly with their basic usages.

In the next chapter, you will be building your first robot – a tabletop fan – and writing your first line of code to make it move!

Further reading

You can refer to the complete part list/inventory of your LEGO BOOST kit at `https://brickset.com/inventories/17101-1`.

You can go through the page at this link and try to remember the original name of each LEGO part in your kit. Consider learning at least 10 names daily and you will be able to remember all the technical terms in just 15 days!

2
Building Your First BOOST Robot – Tabletop Fan

In the summers when we feel hot, a tabletop fan comes in handy! Have you ever come across a situation where you thought of having a portable tabletop fan to beat the summer heat? Let's try to build and code our own portable tabletop fan that we can take anywhere and use!

In this chapter, we will cover the following topics:

- Building the tabletop fan
- Programming the tabletop fan
- Running your tabletop fan at different speeds
- Time for a challenge

By the end of this chapter, you will have a clear idea of how to build basic robots that are similar in complexity to this tabletop fan, as well as completing basic forward-backward programming using the Scratch 3.0 programming language.

Technical requirements

In this chapter, you will need the following:

- A LEGO BOOST kit with six AAA batteries, fully charged
- A laptop/desktop with the Scratch 3.0 programming language installed and an active internet connection

Building the tabletop fan

Before building this fan using your LEGO bricks from the LEGO BOOST kit, let's look at the elements needed to build this fan:

- BOOST Hub
- Motor
- The wings of the fan

Great! We will be building the tabletop fan shown in the following figure:

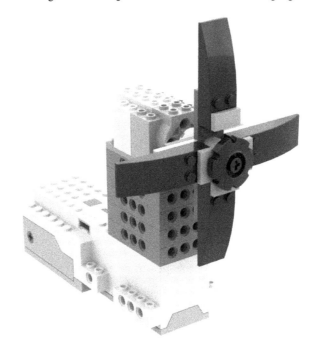

Figure 2.1 – Building a tabletop fan with a LEGO BOOST kit

Follow these steps to build this tabletop fan. Make sure that you select the right pieces from your kit, as mentioned in the building instructions:

1. Take your LEGO BOOST Hub. Make sure that the BOOST Hub has fully charged batteries in it:

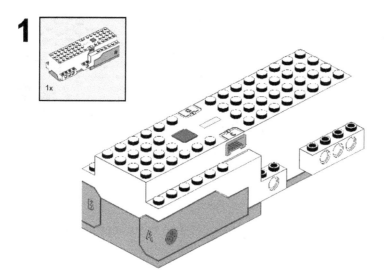

Figure 2.2

2. Take five 4x6 bricks and stack them on the BOOST Hub, as shown here:

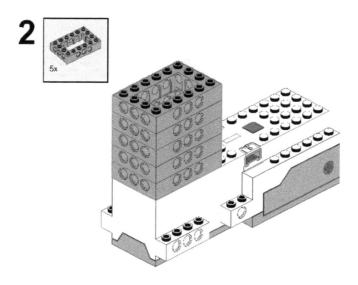

Figure 2.3

3.   Take two 1x6 plates (orange) and place them on the top of the stack:

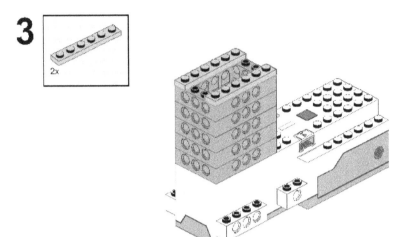

Figure 2.4

4.   Now, take the external motor and mount it on the orange plates that you just attached:

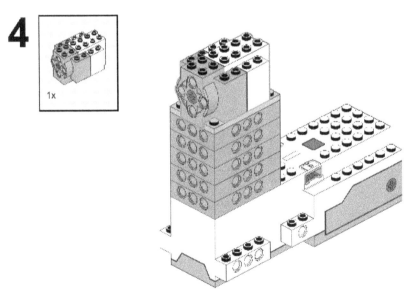

Figure 2.5

5. Now, take a 1/2 bush and one 3M yellow cross axle and attach them to the motor:

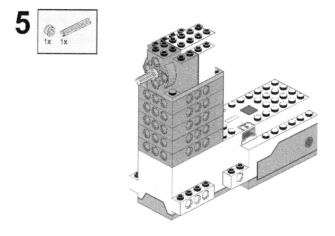

Figure 2.6

6. Take a 2x2 round brick and insert it into the 3M axle:

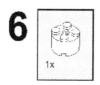

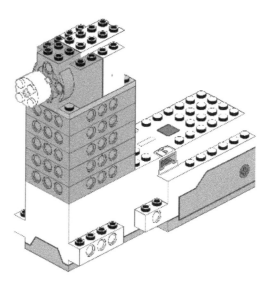

Figure 2.7

7. Take a 2x6 plate and attach it to the white brick with a cross:

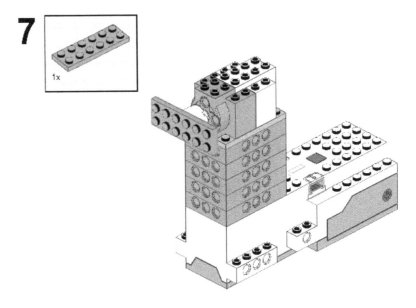

Figure 2.8

8. Take two 2x3 white plates and attach them to the 2x6 blue plate:

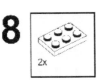

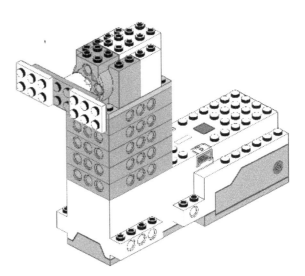

Figure 2.9

9. Take another 2x8 plate and attach it between the two white plates. This will form the base for the fan's wings:

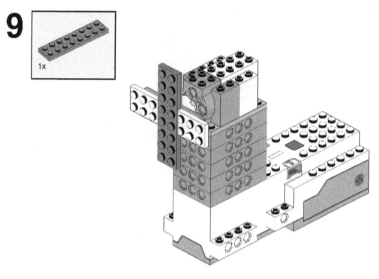

Figure 2.10

10. Take four 2x6 black bricks with bows and attach them to the white and gray plates. These are the wings of your fan:

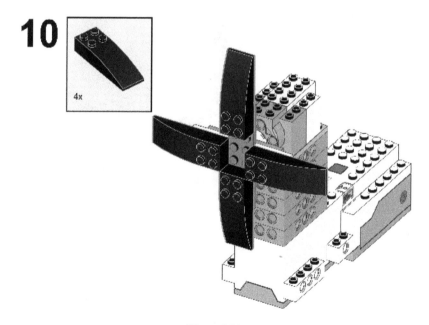

Figure 2.11

11. Take four orange 1x2 plates and attach them to the fan's wings:

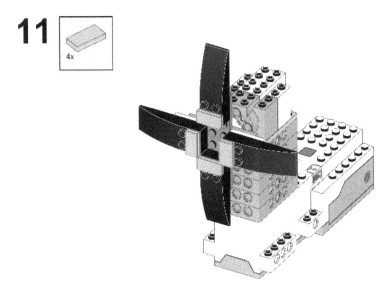

Figure 2.12

12. Take a 2M cross axle and one white brick with a cross. Attach it to the gray plate, as shown here:

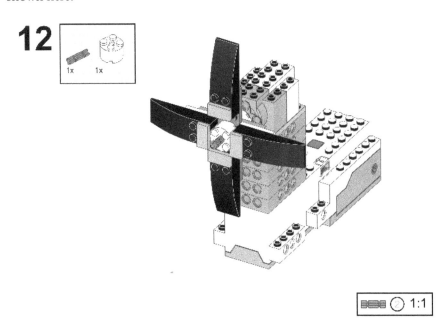

Figure 2.13

13. Take a sprocket and attach it to the 2M cross axle. At this point, your tabletop fan is ready to use! Connect your motor to port C:

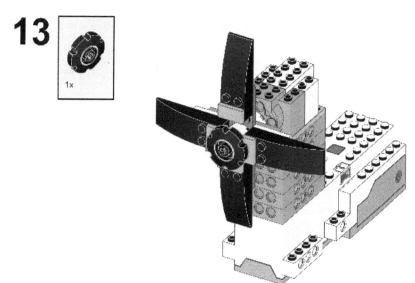

Figure 2.14

Now that your tabletop fan is ready, let's program it.

Programming the tabletop fan

In this section, we will be coding the model we created using the LEGO BOOST kit. There are various ways you can code your BOOST:

- **The LEGO BOOST official app**:

    a. It can only be used on mobile and tablets.

    b. Only the offline version is available.

    c. Only works on Android, Windows, and iOS devices.

    d. It has built-in building instructions for some of the coolest BOOST robots!

    Most of you will be first-time users of this app.

- **LEGO BOOST extension in the Scratch 3.0 programming language**:

    a. It can be used on any device – mobile/tablet/laptop.

    b. It is available as an online as well as an offline version.

    c. It can be used on any OS – Chrome OS/MacOS/Windows/Linux.

    d. The building instructions can be easily and freely downloaded from LEGO Education's official website.

Most of you will have come across the Scratch programming language at least once!

Important Note

Scratch is a block-based programming language developed by MIT Media Labs, US. It is one of the most widely used programming languages by elementary school students across the globe, owing to its user-friendly interface, easy-to-understand programming blocks, and flexibility to connect to various platforms such as BOOST, WeDo, EV3, Arduino, micro:bit, and so on.

Considering these pointers, we shall be using the Scratch 3.0 offline/online editor to code our BOOST robots. Before you begin coding, here are some important instructions you must follow:

1. Before you use the Scratch programming language, you need to install **Scratch Link** on your device, which will help you connect your BOOST kit to the device over Bluetooth. You can download Scratch Link from `https://scratch.mit.edu/boost`.

2. Now, make sure that it is active and appears on your toolbar and that your laptop's Bluetooth is on. Now, let's look at the Scratch programming software.

3. If you wish to use the offline version of Scratch 3.0, you can download it from here: `https://scratch.mit.edu/download`.

4. If you wish to use the online editor for Scratch 3.0, bookmark this link on your device: `https://scratch.mit.edu/`.

5. Once you open this link, click on **Create**. The coding screen will open.

> **Tip**
>
> You may consider signing in on this software for the online version if you wish to save your program and eventually share it with the world, or even use it for yourself in the future!

6. Now that the programming screen is open, click on the **Add Extension** icon, as shown in the following screenshot. This icon has been highlighted with a square here:

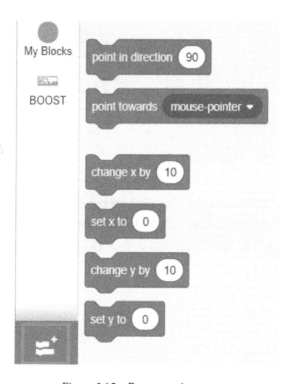

Figure 2.15 – Programming screen

7. Now, select/click on **LEGO BOOST**, as shown in the following screenshot:

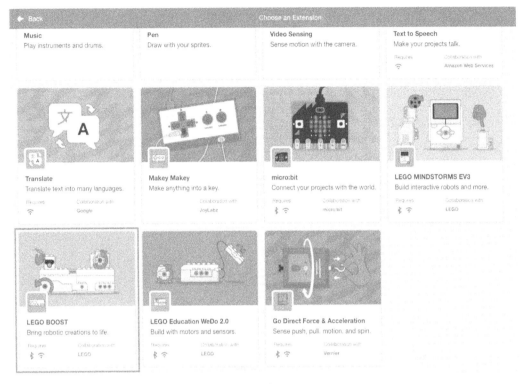

Figure 2.16 – Adding the BOOST extension

8. Now, turn on your BOOST by pressing the blue button. Make sure that it has six AAA-sized battery cells in it. It will start flashing blue the moment you press it.

9. The following screen will appear on your device. Click on **Start Searching**:

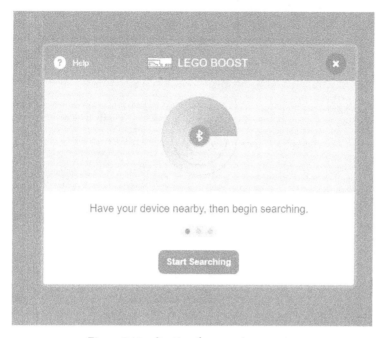

Figure 2.17 – Starting the scanning process

10. Keep your BOOST on and close to the device. The following screen will appear once it has detected your BOOST kit and is connecting:

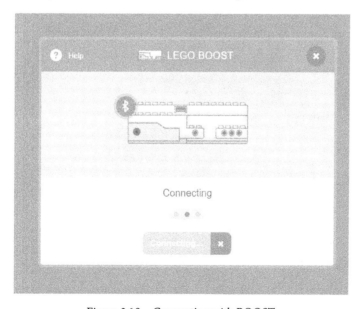

Figure 2.18 – Connecting with BOOST

11. Once connected, the following message will appear:

Figure 2.19 – Connection successful

If you face any difficulties connecting your BOOST Hub to your device via Bluetooth, please visit `https://scratch.mit.edu/boost` for troubleshooting help.

12. The LEGO BOOST programming blocks will be available just below **My Blocks**, under the **BOOST** header:

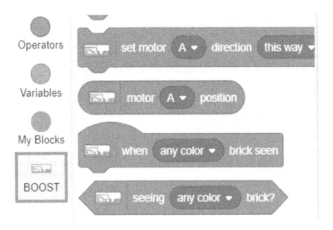

Figure 2.20 – BOOST programming blocks

Voilà! With that, you are all set.

In the next section, we will challenge ourselves by running the tabletop fan you built at different speeds!

Running your tabletop fan at different speeds

Now, let's code our tabletop fan motor (motor C) so that it moves **this way** at 70% speed for 10 seconds when the green flag is clicked.

> **Tip**
>
> Break down your problem statement into small problems before approaching it!

Let's break our problem statement into smaller steps:

- Step 1: Set the event to *when the green flag is clicked*.

- Step 2: Set motor C's speed to 70 percent.

- Step 3: Set motor C's direction to **this way**.

- Step 4: Turn motor C on for 10 seconds.

- Step 5: Turn motor C off.

If you break down your problem statement like this and write these step-by-step instructions each time, your coding job will be extremely easy and fun! Since this is your first time coding, please use the following sample code for this task:

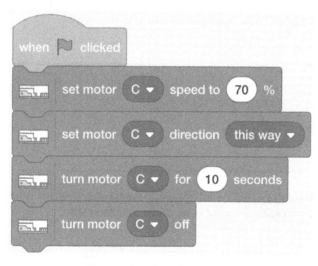

Figure 2.21 – Sample program

Test it and see how much fun it is to use your fan to beat the summer blues!

Now, can you write some code that will move your tabletop fan motor C **that way** at 50 percent power for 10 seconds when the green flag is clicked?

Write down your problem steps here:

- Step 1: _____
- Step 2: _____
- Step 3: _____
- Step 4: _____
- Step 5: _____

Now, write your code and see how it works! Identify any errors and rectify them. Make sure that your code works exactly the way we want it to work!

Now that you know how to run your tabletop fan, the next section will test your skills.

Time for a challenge

Now, it's time to test your learning! Read the following problem statement twice:

P1: Code your tabletop fan to move **this way** at 80 percent power for 5 seconds, then move **this way** at 50 percent power for 5 seconds, and then stop!

Problem fragmentation:

```




```

P2: Code your tabletop fan to move **that way** at 70 percent power for 6 seconds, then move **that way** at 30 percent power for 3 seconds, and then stop!

Problem fragmentation:

```




```

Use your imagination and use the sound and display blocks creatively to make your project more interactive and fun! For example, you can display the power that your tabletop fan is currently moving at, use sound blocks that replicate the humming sound of a fan, or anything of your choice!

> **Tip**
>
> You can find the display and sound blocks in the *looks* and *sound* programming pallets, respectively.

Remember that the sky is the limit due to the tools and the coding platform that you have at your disposal.

> **Fun Fact**
>
> Did you know that the tabletop fan was invented by *Schuyler Skaats Wheeler* in 1882?

Summary

In this chapter, you built your first robot using the LEGO BOOST kit: a tabletop fan. Your excitement level was taken one step up by us introducing coding in this chapter. You also learned how to add the BOOST extension to the Scratch 3.0 programming language and connect your BOOST to your device using Bluetooth. By completing a basic exercise, you explored the basics of programming, including turning the motor in a certain way, setting the motor's power, and exploring blocks from the looks and sound pallets. With this basic knowledge, you should be able to build some basic robots using your BOOST kit and write some code to make them move.

In the next chapter, we will be building a robot that will be able to move forward and backward without wheels! Exciting, isn't it?

Further reading

To enhance your Scratch programming skills, go to `https://scratch.mit.edu/projects/editor/?tutorial=all` and build some cool games and animations. This will help you enhance your coding skills, which you will use in the upcoming chapters of this book.

3
Moving Forward/Backward Without Wheels

I am sure that building and coding the table-top fan in the previous chapter was fun for you. In this chapter, we will be building a robot that can move without wheels both forward and backward. Have you ever thought about how to make things move from one place to another without wheels? Hint – it is there in your surroundings itself. Try to answer these questions:

1. How do we, human beings, move? _____

2. How does a snake move? _____

3. How does a frog move? _____

There are other forms of movement as well, such as wiggling, jiggling, creep, leap, and bouncing, which we will explore later. Walking, hopping, slithering, and rolling are some of the most common locomotion techniques used by robots today, and we shall learn about these forms of movement in detail in this chapter.

In this chapter, we shall focus on the core robotics model and eventually increase the complexity of programming as well. In this chapter, you will build a robot that moves with the help of legs and linkages. Furthermore, you will strengthen your construction skills by building this robot using your LEGO BOOST kit.

In this chapter, we will cover the following topics:

- Building a robot without wheels
- Let's code the robot without wheels
- Time for a challenge

We will end the chapter with some captivating challenges for you to solve. Let's discuss types of movement first.

Different types of movement

Let's first try to understand the various ways in which we can make our robot move. These movements are like some animals' locomotion methods:

- **Walk** – This type of movement is the human method of moving around. Observe the movement of your legs when you move. They have various linkages, such as the knees, ankles, toes, and so on. All these things in coordination ensure that we can move smoothly across most types of terrain. Can you name at least three animals that use walking as their mode of movement? _____

- **Hop** – Hopping is basically jumping on either both your feet or on one foot. Animals such as frogs, kangaroos, and grasshoppers hop! MIT started developing hopping robots in the 1980s.

- **Slither** – Snakes slither. Robots can also move in this way. The Japanese **ACM-R5** snake robot can navigate both on land and water.

- **Rolling** – Any robot that uses wheels to move is a rolling robot. They are most efficient on smooth surfaces.

Let's move on to building this robot now.

Technical requirements

In this chapter, you will need the following:

- LEGO BOOST kit with 6 AAA batteries, fully charged
- Laptop/desktop with Scratch 3.0 programming and an internet connection
- A ruler, pencil, and paper, along with some small, colorful sticky notes

Building a robot without wheels

In this section, we are going to build a robot that can hop, as shown in the following image:

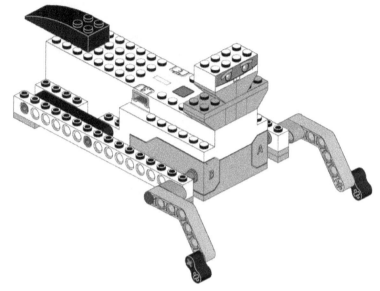

Figure 3.1

Let's build the robot by following these building instructions:

1. Take your BOOST Hub. Ensure that all six batteries are fully charged:

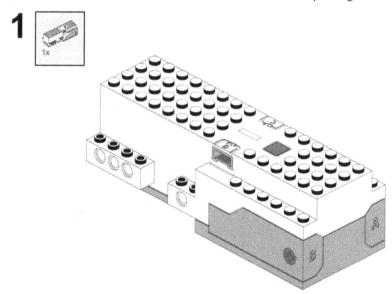

Figure 3.2

2. Take two 3M pegs and connect them with the BOOST Hub:

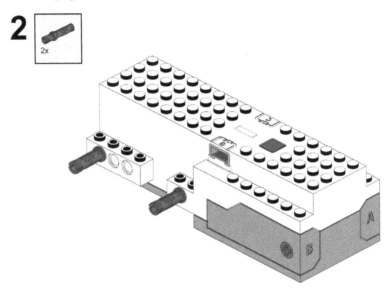

Figure 3.3

3. Now take a 7M beam and connect it to the pegs that you just attached to the BOOST Hub:

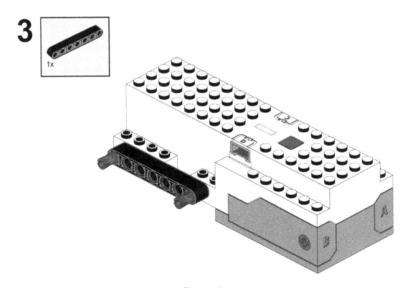

Figure 3.4

4. Now take a 16M Technic beam and connect it to the 7M beam:

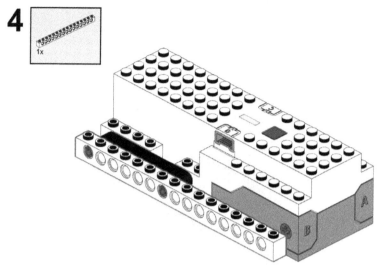

Figure 3.5

5. Now, take a full bushing, a 1x2 beam with cross and hole, a 5M cross axle with an end stop, and a 4x4 Technic angular beam. Take this angular beam and the 1x2 beam and pass your 5M cross axle through this. Make sure that this axle passes through the cross of the 1x2 beam. Now, place the full bushing between the hub and the Technic beam and push the axle in until it is connected with motor port B:

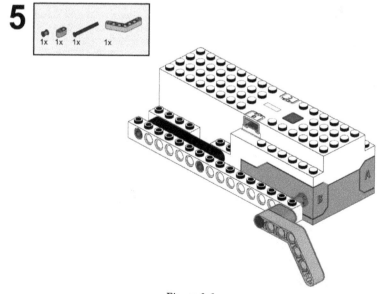

Figure 3.6

6. Let's build a similar structure on the other side. Turn the BOOST Hub around. Take two 3M pegs and connect them to the hub as shown:

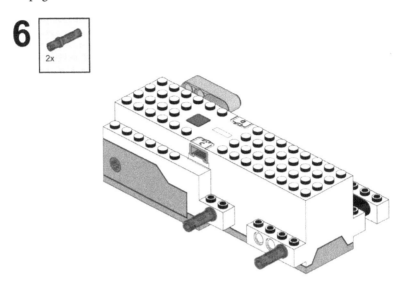

Figure 3.7

7. Now, take a 7M beam and connect it to the pegs:

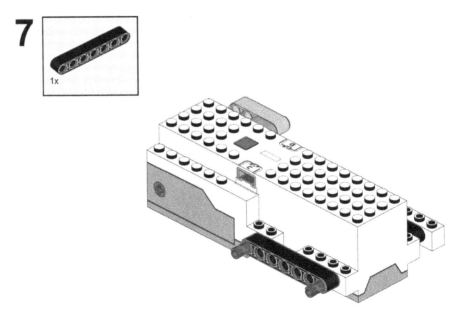

Figure 3.8

8. Now take a 16M technic beam and connect it to the 7M beam:

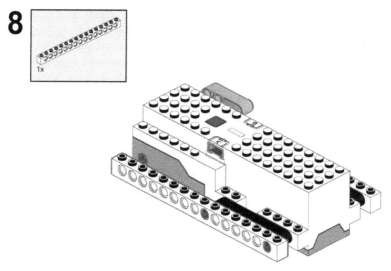

Figure 3.9

9. Now take a full bushing, a 1x2 beam with cross and hole, a 5M cross axle with end stop, and a 4x4 Technic angular beam. Take this angular beam and the 1x2 beam and pass your 5M cross axle through it. Make sure that this axle passes through the cross of the 1x2 beam. Now, place the full bushing between the hub and the Technic beam and push the axle in until it is connected to motor port A:

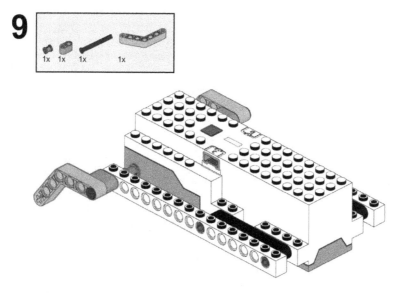

Figure 3.10

10. Now, flip your BOOST Hub upside down. Take two 2x3 plates and connect them to the BOOST Hub:

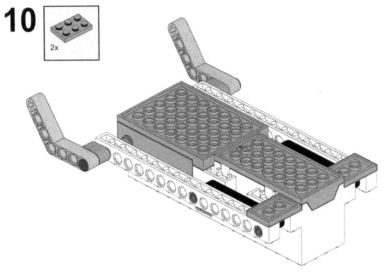

Figure 3.11

11. Now take four 1x2 plates, a 2M axle, and one damper. Stack two 1x2 plates, such that they make a pair. Connect them on both sides of the hub as shown. Also, connect the damper to the 2M axle and connect this axle to the cross hole of the 4x4 angular beam:

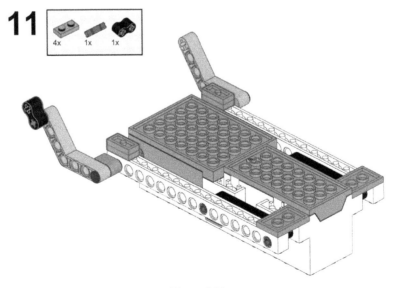

Figure 3.12

12. Take a damper and a 2M axle. Connect them to the other leg with the cross hole of the 4x4 angular beam:

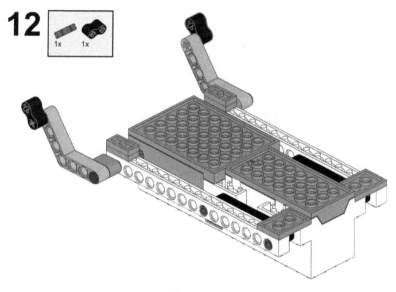

Figure 3.13

13. Get your BOOST Hub in the upright position again. Take a 2x6 brick with a bow and attach it to the hub:

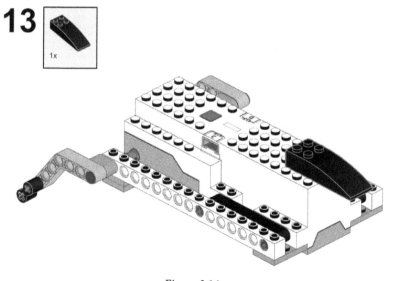

Figure 3.14

14. Take two 2x2 roof tiles and connect them to the hub as shown:

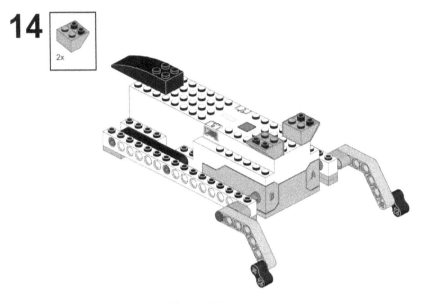

Figure 3.15

15. Now take two 1x2 bricks and connect them in between the roof tiles on the BOOST Hub:

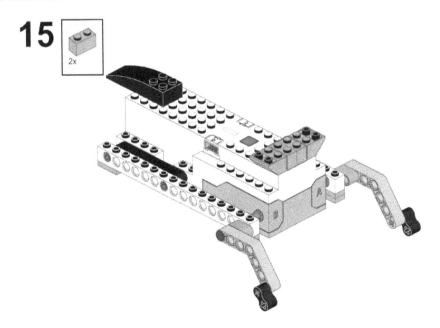

Figure 3.16

16. Take a 2x6 plate and connect it to the 1x2 bricks and roof tile bricks as shown:

Figure 3.17

17. Take two 2x2 roof tiles and connect them to the 2x6 plate that you just attached:

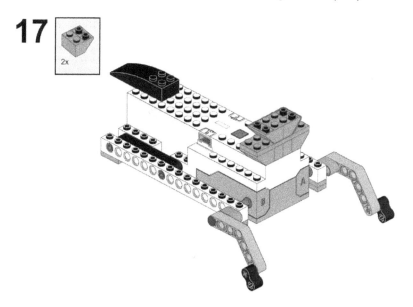

Figure 3.18

18. Now take a 2x4 white brick with an eye design and connect it to the 2x2 roof tiles:

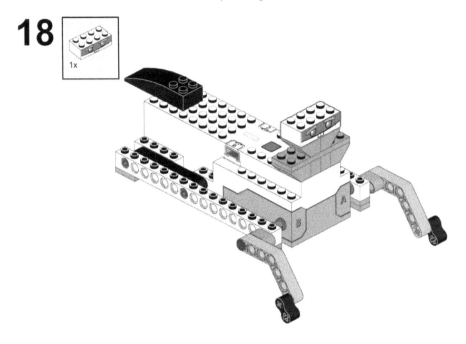

Figure 3.19

The robot that you have built moves in a similar way to a frog: hopping. It will be fun to write a program for this hopper bot.

Let's code the robot without wheels

Open Scratch 3.0 and connect your BOOST Hub to your laptop as per the instructions given in *Chapter 2*, *Building Your First BOOST Robot – Tabletop Fan*. While performing some activities today, we shall understand the difference between seconds-based and rotation-based programming.

Before we move on to the task, let's learn about two new programming blocks:

- **set light color to**: This block changes the light color of your BOOST Hub:

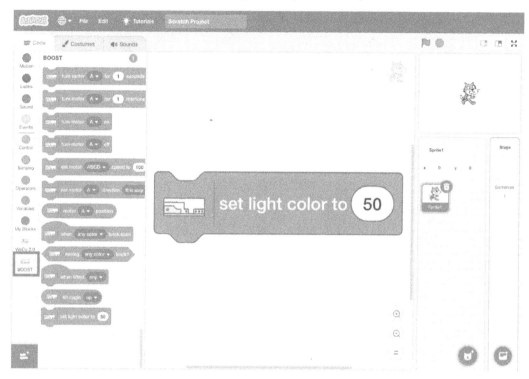

Figure 3.20 – New block to learn

- **wait**: This block waits for the number of seconds that the user inputs and lets the previous block execute for that time. We will learn how to use it while writing some code in the tasks. This block is found in the **Control** pallet:

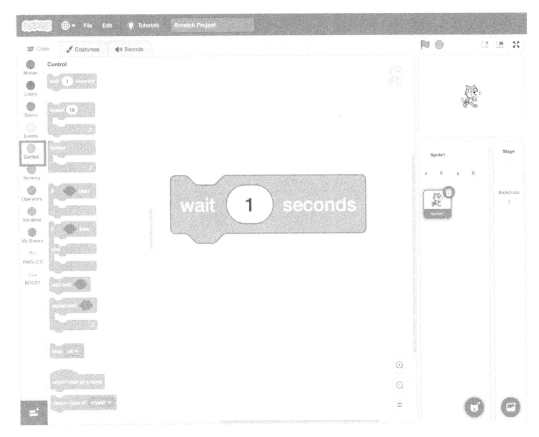

Figure 3.21 – Wait block in the Control pallet

Set light color:

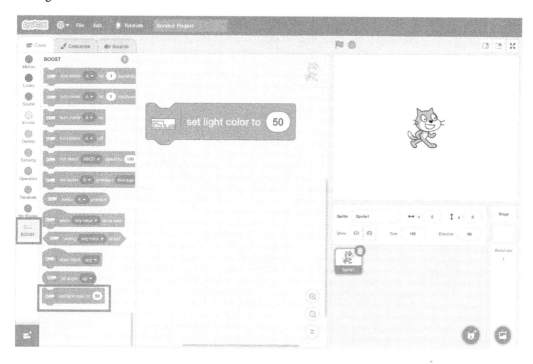

Figure 3.22 - Changing the hub light colors

This is how the light color changes based on the value that you input:

Input value in set light color to	Light color on Lego BOOST kit
100	Red
80	Purple
60	Blue
40	Light green
20	Green
10	Yellowish green

Table 3.1 – Hub light colors

I hope you remember the following tip – break down your problem statement into smaller problems.

Activity #1

Program your robot to move forward at 40% speed for 5 seconds. Mark the start (green sticky note) and stop (red sticky note) positions of your robot. Change the hub light color to *blue* for 2 seconds when the robot stops.

Let's break our problem statement into smaller steps:

1. Set the event to **when green flag is clicked**.

2. Set the **motor AB** speed to 40%.

3. Set the **motor AB** direction to **this way**.

4. Turn on **motor AB** for 5 seconds.

5. Set light color to 60.

6. Wait for 2 seconds:

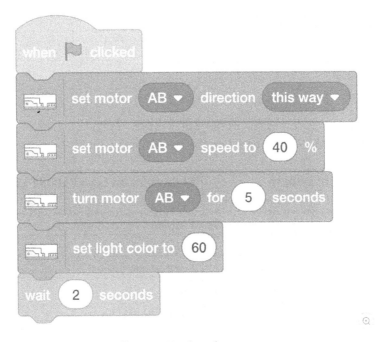

Figure 3.23 – Sample program

Great! Your program is now ready to be downloaded to your BOOST Hub and run. See how your robot hops with this code.

Activity #2

Program your robot to move forward at 40% speed for 5 rotations. Make sure that your starting position is the same as in *Activity #1*. Mark the stopping position with an orange or any other color sticky note. Change the hub light color to *green* for 2 seconds when the robot stops.

Write down your problem fragmentation here:

- Step 1 - _____
- Step 2 - _____
- Step 3 - _____
- Step 4 - _____
- Step 5 - _____
- Step 6 - _____
- Step 7 - _____

Once you've done this, write your code in Scratch and run the robot.

Did you observe the difference between the two activities? Although your starting position and motor power are the same, your stopping position is different in each activity. It means that *move for __ seconds* and *move for __ rotations* are completely different.

Rotation-based movement gives precision to our robot's movement. Regardless of the battery level, your robot will always travel the same distance in rotation-based programming.

But in seconds-based programming, the distance traveled by your robot will be dependent on the battery level of your BOOST Hub. If the battery level is low, the distance traveled for the same time and at the same motor power will be less than if the battery was at full power. This reduces the consistency of your robot's performance. There are some advantages associated with seconds-based programming, which you will explore in the chapters coming up next.

Activity #3

Since you have marked the starting and stopping position of the robot in *Activity #1* and *Activity #2*, let's use the ruler and measure the distance covered by your robot in each case:

Distance covered in activity #1 = _____ cm

Distance covered in activity #2 = _____ cm

As you can see by your answers, it is not true that the robot moves the same distance when it goes forward for either 5 seconds or 5 rotations. They are different.

Time for a challenge

Time to test your learning! Read the problem statement twice.

Challenge #1

Do you know that you can measure the speed of your robot if you have its distance traveled as well as the time taken? The formula to calculate speed is *total distance traveled / total time taken*. Try to calculate the speed of your robot in each case. You can use the empty spaces in the following sections to do the math work.

Distance traveled in *Activity #1* = _____ cm

Time taken = 5 seconds

Speed = _____ cm/second

Distance traveled in *Activity #2* = _____ cm

Time taken = _____ seconds (run your robot again and measure the time taken)

Speed = _____ cm/second

Challenge #2

Program your robot to move forward at 80% speed for 5 seconds. Mark the start and end positions and calculate the speed of your robot.

Distance traveled = _____ cm

Time taken = 5 seconds

Speed = _____ cm/second

```
┌─────────────────────────────────────────────────────────────────────┐
│                                                                       │
│                                                                       │
│                                                                       │
│                                                                       │
│                                                                       │
│                                                                       │
└─────────────────────────────────────────────────────────────────────┘
```

Did you observe that your robot has traveled more distance in the same time of 5 seconds than in *Challenge #1*? In seconds-based programming, the distance traveled by the robot is purely based on the motor power and the amount of time the motor runs for. So, for the same period, the robot travels further if motor power is high, and it travels a shorter distance if the motor power is low.

Challenge #3

Program your robot to move forward at 80% speed for 4 rotations. Mark the start and end positions and calculate the speed of your robot.

Distance traveled = _____ cm

Time taken = _____ seconds

Speed = _____ cm/second

Did you observe that your robot has covered the same distance as in *Challenge #2* but in less time because you increased the motor speed level? Rotation-based programming always adds consistency to your robot's performance.

Summary

In this chapter, you learned how to build a robot that can move forward and backward without wheels. You also learned about various ways in which things can move even without wheels, such as by hopping or slithering. A new programming block called **brick light** was introduced, which can change the light color of the BOOST Hub brick. You also learned the difference between rotation-based and seconds-based programming, with their respective advantages and disadvantages. In the next chapter, you will learn how to build a basic robot with wheels and learn more about programming, making your robot turn, and making shapes such as squares, circles, and triangles.

Further reading

Do you know that satellites are objects launched by humans into orbit for various purposes? They revolve around the Earth, collect data as they are programmed to, and send the data back to Earth. This data is then used for various purposes. Google Maps and Apple Maps are dependent on data sent by satellites. You can read more about satellites at `https://www.esa.int/kids/en/learn/Technology/Useful_space/Satellites`.

4
LEGO BOOST Rover

Anything that reduces human effort is called a machine. A simple machine is a mechanical device for applying force. Any complex machine is a combination of two or more simple machines. We will learn how to apply a few of these simple machines in our upcoming projects. In science, there are a total of six simple machines, as follows:

- Wheel and axle
- Pulley
- Lever
- Wedge
- Screw
- Inclined plane

Wheels and axles help move heavy loads from one place to another by reducing the friction between the load and the surface, owing to its round shape. The wheel is considered one of the greatest discoveries of mankind. You can find numerous examples in and around your vicinity where wheels are used extensively to make things move easily, for example, your holiday suitcase! Have you ever tried picking up your suitcase when it is completely full? It is extremely heavy and almost impossible to lift, right? And what does your suitcase have? It has four small wheels at the bottom that makes moving such a heavily loaded suitcase extremely easy and fun. You can find many such applications of wheels and axles in your day-to-day life. Can you think of at least three places where you have seen this application of wheels and axles?

1. _____

2. _____

3. _____

Great! In this chapter, you will make use of wheels and axles to help the robot move freely on the floor. You will also learn about two different types of turns that a robot can perform and execute those turns in the respective activities. Exciting, isn't it?

In this chapter, we will cover the following topics:

- Building the BOOST rover

- Making the rover move

- Time for a challenge

Technical requirements

In this chapter, you will need the following:

- A LEGO BOOST kit with six AAA batteries, fully charged

- A laptop/desktop with Scratch 3.0 installed on it and an active internet connection

- An A3 size drawing sheet with a pencil and ruler

Building the BOOST rover

In this section we will build the robot as shown in the following figure:

Figure 4.1

Follow the given steps to build the robot:

1. Make sure you insert six AAA working batteries into your BOOST Hub before you start the construction process:

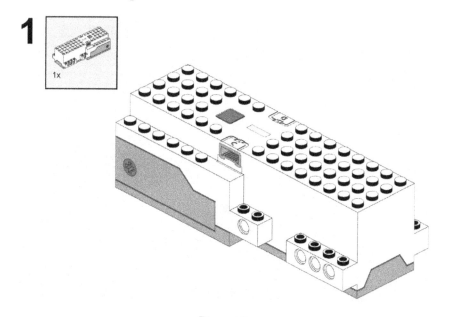

Figure 4.2

2.   Now, take two 2x6 bricks with bows. Attach them to the BOOST Hub:

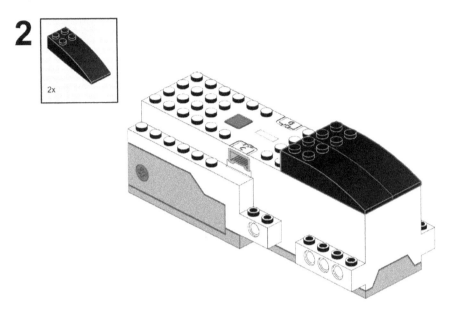

Figure 4.3

3.   Now, take two 1x2 plates with ball cups. Connect them to the 2x6 bricks with bows:

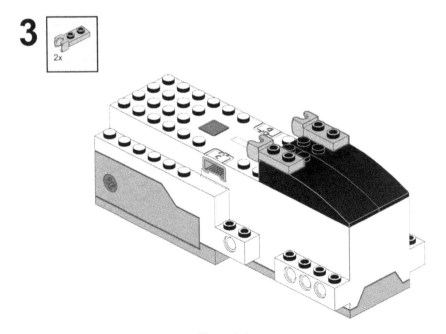

Figure 4.4

4. Take a 4x6 brick and connect it to the BOOST Hub, as shown here:

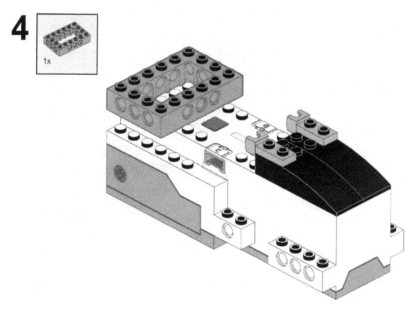

Figure 4.5

5. Take a 2x4 brick, a 2x6 plate, and a 2x4 bearing element with a snap. Attach the 2x6 plate first and then attach the 2x4 brick to it. Now, connect this 2x4 bearing element with a snap on the top of it:

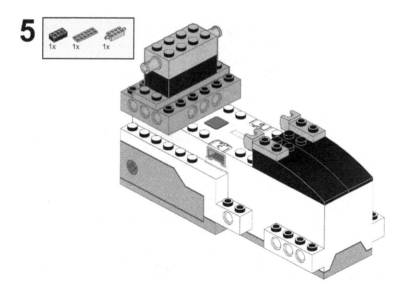

Figure 4.6

6. Take two 5.5M axles with 1M stops and connect them to both motor A and motor B of the BOOST Hub:

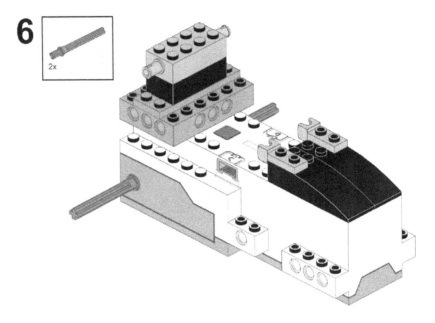

Figure 4.7

7. Now, take two 3x5 angular beams and connect them to the axle on both sides:

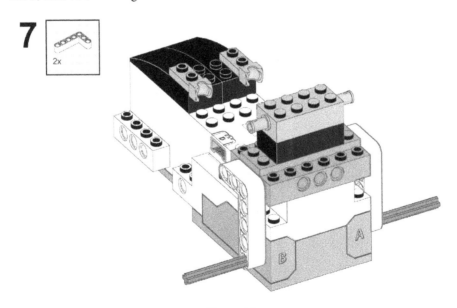

Figure 4.8

8. Take four 2M friction snaps with cross holes. Connect them to the 3x5 angular beam on both sides via the first and third holes:

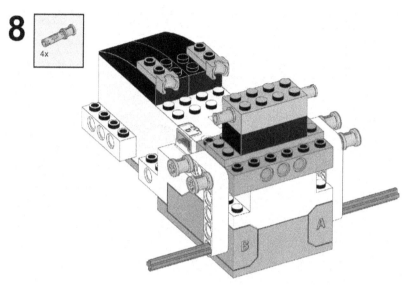

Figure 4.9

9. Now, take two 2x2 round bricks with crosses and two 2x2 flat round tiles with a hole in it. Connect them to the axles on both sides. Ensure that the studs of the round bricks with crosses are attached firmly to the flat round tiles:

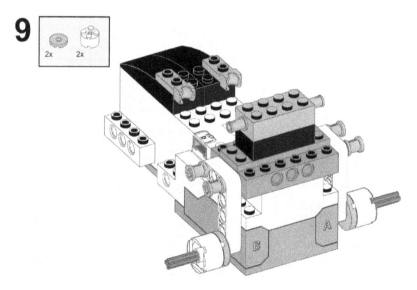

Figure 4.10

10. Now, take two wheels and connect them to the axle with the connection that you just made. This wheel must slip over the round brick and plate:

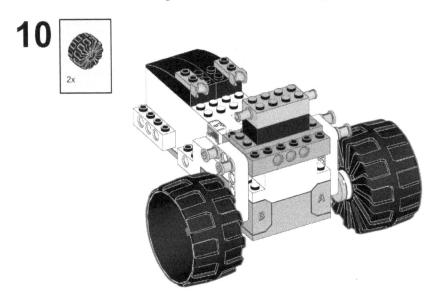

Figure 4.11

11. Now, connect two black pegs to both sides of the BOOST Hub:

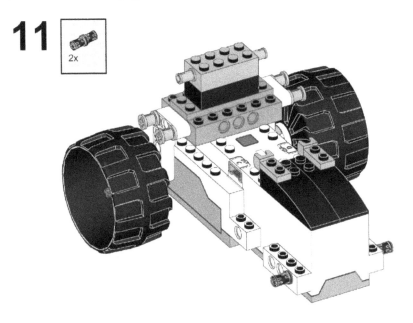

Figure 4.12

12. Now take two 3x5 angular beams and connect them to the pegs that you just attached to the hub. Here, we are building the driven wheel:

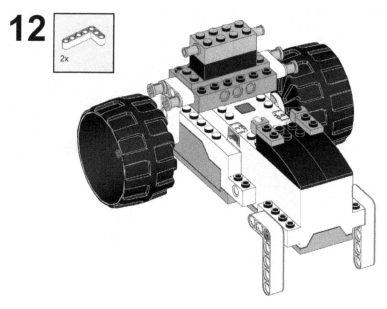

Figure 4.13

13. Take two friction snaps with crosses and connect them to the 3x5 angular beam:

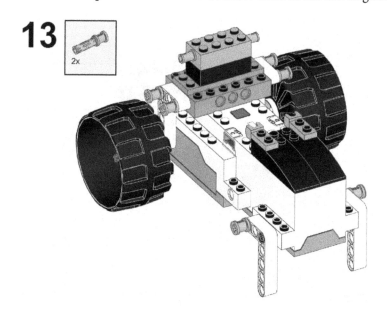

Figure 4.14

14. Take four black pegs and connect them to the 3x5 angular beam from the inside:

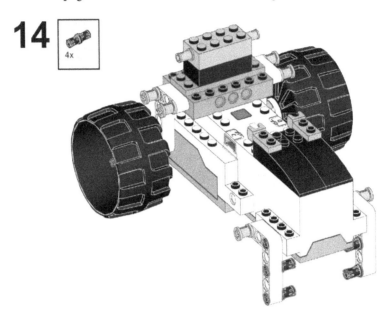

Figure 4.15

15. Take two 9M beams and connect them to the recently attached pegs on the angular beams on both sides of the robot:

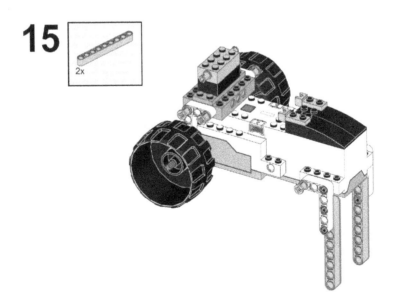

Figure 4.16

16. Take two black pegs and two connector pegs with cross axles. Attach them to the 9M beams via the holes shown in the following figure:

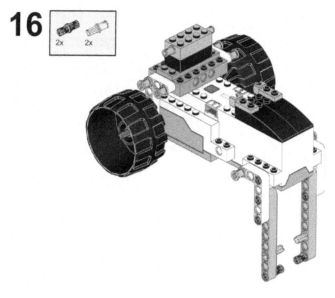

Figure 4.17

17. Take two 4x4 angular beams and connect them to the 9M axles, as shown here. Make sure that the cross holes of the 4x4 angular beams are connected to the connector pegs of the cross axles:

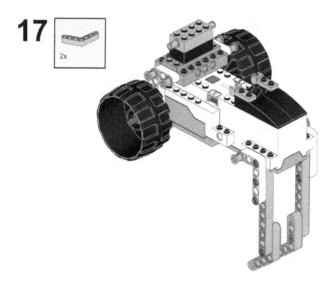

Figure 4.18

18. Take two hub wheels and a 4M cross axle. Connect them to the angular beam by its last cross hole. This connection will serve as your driven wheel:

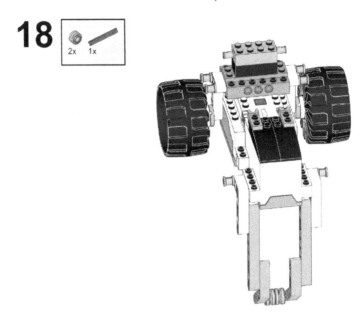

Figure 4.19

19. Take two 5M beams and connect them to the 2x4 brick with snaps:

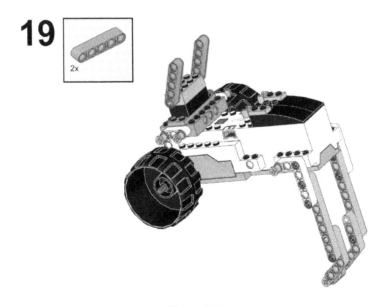

Figure 4.20

20. Now, take two 1x2 bricks with snaps and connect them to the other ends of the 5M beams that you just attached:

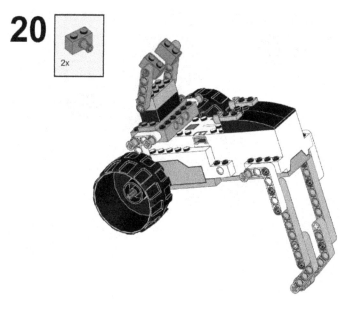

Figure 4.21

21. Take a 1x4 plate and a 2x4 plate. Connect them to the 1x2 bricks with snaps, as shown in the following figure:

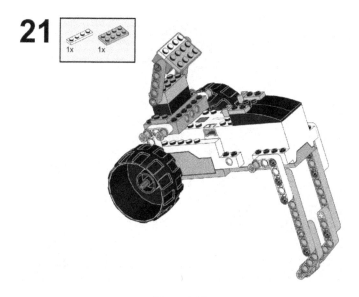

Figure 4.22

22. Take two 4x4 plates that are 1/4 circles. Connect them to the underside of the 2x4 plate that you attached in the previous step:

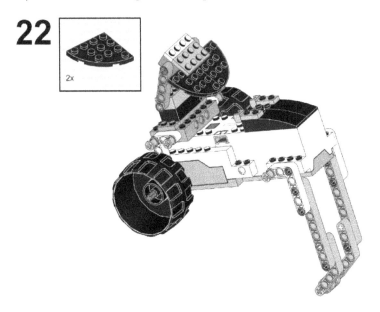

Figure 4.23

23. Take a left 3x8 angle plate and a right 3x8 angle plate and connect them to the recently connected LEGO bricks:

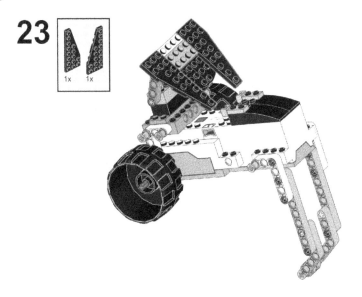

Figure 4.24

24. Take two 1x8 flat tiles and use them to connect the left and right 3x8 angle plates, along with the other LEGO bricks. Then, take a 1x10 brick and connect it so that it is in between the two flat tiles. This will make this connection stable:

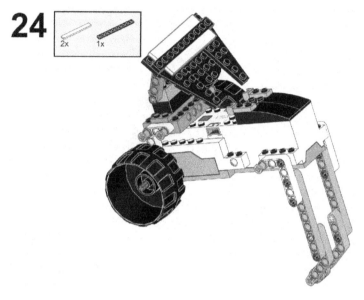

Figure 4.25

25. Take a 4x4 LEGO plate and place it on top of the 1/4 circle plates:

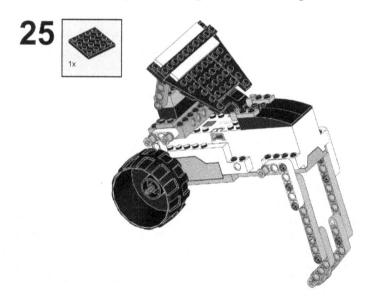

Figure 4.26

26. Take two 1x2 plates with ball ends. Connect them with the already attached 1x2 plates with ball cups from one end and on the 3x8 plates with angle on the other end. Now, take two 4x4 plates with arc and connect them as shown in the following figure:

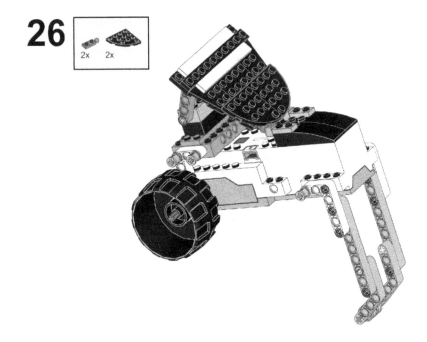

Figure 4.27

The robot that you have built here uses the two internal motors of your BOOST Hub to power two wheels individually. Your turning becomes possible and easy when your drive wheels are powered individually. Remember one thing – the speed difference between your two drive wheels makes turning possible.

Making the rover move

First, let's learn how to make our robot move straight by powering both motors at the same time. We will be using the same programming blocks that we used in *Chapter 2, Building Your First BOOST Robot – Tabletop Fan*, and *Chapter 3, Moving Forward/Backward Without Wheels*, but we will be selecting the **AB** option from the drop-down menu, as shown in the following screenshot. Here, I have written a program to make the robot go forward for two rotations at 100% speed:

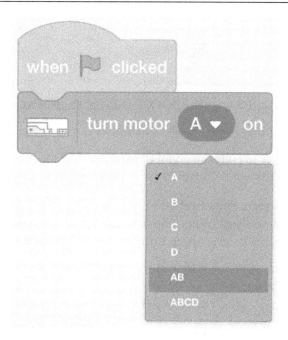

Figure 4.28 – Selecting "AB" from the drop-down menu

Now that you know how to power both motors at the same time, let's write the main code:

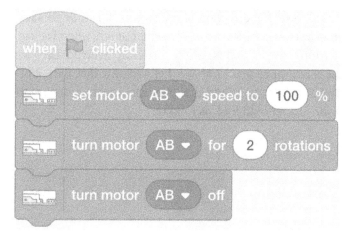

Figure 4.29 – Sample program for forward programming

Now, let's perform a few activities.

Activity 1

Can you program your robot to do the following?

1. Move forward for 2 seconds at 60% speed.

2. Wait for 2 seconds and change the BOOST Hub light's color to purple.

3. Move backward for three rotations at 30% speed.

Great! Now, let's learn how to code our robot to follow an *L* shape. For this, we need to understand how to turn the robot. There are two different types of turns:

*   **Point turn**: In this turn, one wheel moves forward while the other wheel moves backward at the same speed. When you perform this action, your robot turns on its own point/position. Wherever sharp turns are needed, point turn is used.

Before you program your robot to turn, you need to learn about one new block; that is, **set motor direction**:

Figure 4.30 – Set motor block

This block is used to set the direction of the motors. It can be **this way** (clockwise), **that way** (anti-clockwise), or **Reverse**. We will be smart when we use these blocks in our coding from now on; that is, where turns are involved. Let's write some code that makes the robot move in an *L* shape. Let's assume that the robot will move forward for two rotations, turn, and then move forward for one rotation to draw an *L* shape. Your code should look as follows:

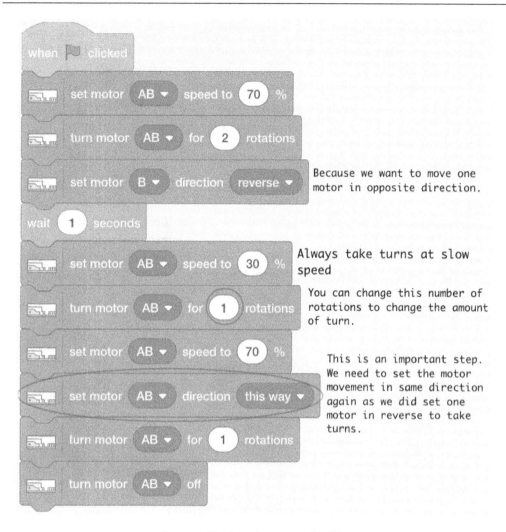

Because we want to move one
motor in opposite direction.

Always take turns at slow
speed

You can change this number of
rotations to change the amount
of turn.

This is an important step.
We need to set the motor
movement in same direction
again as we did set one
motor in reverse to take
turns.

Figure 4.31 – Sample program for "L"

Which direction did your robot turn? Left, or right? _____ (Space has been
left for you to answer.)

Do you know that you can change the motor that you reversed before making this turn to change the turn's direction? Just try it and see how this works. Basically, if you want to turn right, your right wheel should move backward while your left wheel should move forward, and vice versa for the left turn. Also, you need to change the number of rotations that your motors move for while turning to achieve the exact amount of turn that you want. You can set your values in numbers with decimals. Two rotations are fine, but you can also set 2 . 1, 2 . 2, or even 2 . 3 rotations so that you make precise turns every time.

- **Swing turn**: In this turn, one wheel moves, while the other wheel stands still. Here, whichever direction you want to turn in, the wheel on that side stands still. So, for example, if you want to turn right, the right hand-side wheel stays still and only the let hand-side wheel moves. This forms an arc-like movement where the turns are wide. Whenever you need to make wide turns, you should use swing turns. Write a program so that the robot moves forward for 2 seconds and then takes a swing turn using motor A for 1 second:

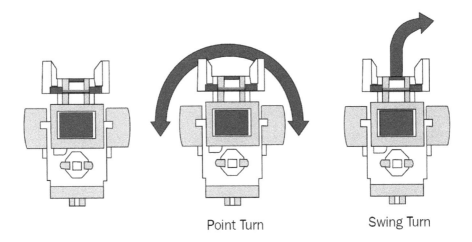

Point Turn   Swing Turn

Figure 4.32 – Different types of turns

So, which direction did your robot turn? Left, or right? _____ (Space has been left for you to answer.) As the name suggests, the robot swings on one wheel to make the appropriate turns.

Did you notice the stark difference between the point and swing turns here? You will apply these turns based on your needs in the upcoming activities.

Activity 2

Program your robot to move in a square, as shown in the following diagram:

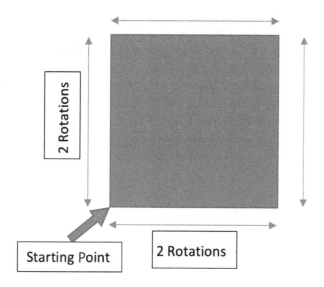

Figure 4.33 – Square shape for reference with measurements

Let's write an algorithm for this code:

1. Move forward for two rotations.
2. Turn right.
3. Move forward for two rotations.
4. Turn right.
5. Move forward for two rotations.
6. Turn right.
7. Move forward for two rotations.
8. Turn right.

Which turn will you be using here – point or swing? Since you need sharp turns at the edges of the square, you must use point turns for this activity. Follow the algorithm and write some code to make your rover move in a square shape. Did you observe something irritating in this code? You are writing the same line of code four times to form a square, hence making this code iterative. Your robot must follow the same action four times – move forward and turn right. How can we reduce these iterations and write fewer lines in the code for the same task? For such repetitive actions, we can use loops. In loop-based programming, you can write basic code that then needs to be followed *n* times. In the case of this square, you need to repeat an action four times. In such cases where the number of repetitions is specific, you can use a block called **repeat** from the **Control** pallet area of Scratch. Let's see how this works! Write the following code on your device:

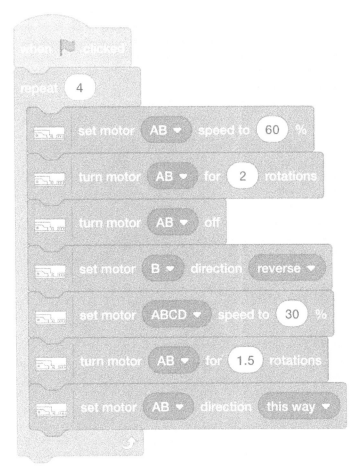

Figure 4.34 – Square shape with repetition applied

Just adjust the following parameters before running this code to form a proper square:

- Number of rotations to turn for

- Motor set in reverse

Similarly, when you want to repeat some actions forever, you must use the **forever** block from the **Control** pallet area:

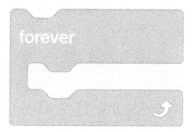

Figure 4.35 – Forever loop

Try using this forever loop instead of the repeat-based loop in the same code. See what happens. Your robot will constantly move in a square shape until it is stopped by the device. Now that you are good at taking turns and using loop blocks, you may wish to write some code that makes your robot move in a rectangle shape twice:

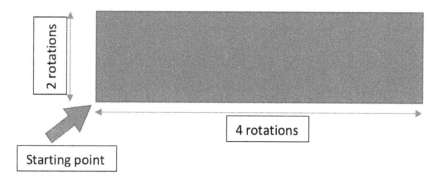

Figure 4.36 – Dimensions for the rectangle shape

It looks like you have got the hang of coding your robot to make it move in shapes with accurate turns. Try to code your robot to move in shapes such as triangles, hexagons, and diamonds with the dimensions of your choice.

Once you've done this, you can create your own path with a few left and right turns in it – some sharp, some wide – and try to run your robot through that path. This will enhance your coding skills with your BOOST kit.

Time for a challenge

Challenge #1

One more way to make turns is by simply running both motors at different speeds. By doing so, your robot will steer just like your car, instead of turning sharp at that very point. This technique can be useful when you want to make smooth, long turns of form shapes such as an arc or a circle. Either draw a big circle or find a big enough circular object in your home and try to code your rover to move around it using this technique.

Challenge #2

Can you build a seesaw using the external motor and LEGO parts in your BOOST kit? A seesaw is a common example of a simple machine lever.

Summary

In this chapter, you learned how to use wheels and axles. You applied these to your robot and built your own rover with wheels using the BOOST kit. You also learned about new programming blocks such as set motor direction, repeat loop, and forever loop. We also looked at two different types of turns that are commonly used when coding robots to make them turn; that is, point and swing turns.

In the next chapter, you will be introduced to the fascinating world of gears and how to apply them to build a geared robot. You will perform a few activities to understand concepts such as gear up and gear down. This is going to be a lot of fun!

Further reading

As we mentioned at the beginning of this chapter, there are six simple machines you can use. We covered wheels and axles in this chapter. You should learn about the other five simple machines as well by going to the following page:

```
https://www.britannica.com/technology/simple-machine
```

If you know how to practically apply these simple machines, you will be able to build your own mechanisms in the future.

5
Getting into Gear – My First Geared Robot

A wheel with teeth on it is called a gear. A gear is used for various applications, including the following:

- Changing the speed of an object
- Changing the direction of the force that's being applied
- Changing the torque of an object

Whether it is your parents' car or their bike, vehicles use gears to move. Gears are used in almost all machines, mechanisms, and robots. You can find at least one application of gears in your printer, machine tools such as drills, industrial robots, packaging machines, and so on. The most common example of where you can find gears is in the wall clock in your home! The hours, minutes, and seconds hands are all interconnected with the help of complex gear mechanisms.

In this chapter, you will learn about various concepts surrounding gears, such as gear ratios and torque, by performing some useful hands-on experiments. We will be covering the following topics:

- Using different types of gears based on your requirements
- Understanding some important terminologies
- Building the geared robot
- Let's code the robot in gear up and gear down mechanism
- Time for a challenge

First, let's build the geared robot.

Technical requirements

In this chapter, you will need the following:

- A LEGO BOOST kit with six AAA batteries, fully charged
- A laptop/desktop with Scratch 3.0 installed on it and an active internet connection
- A diary/notebook with a pencil and eraser

Using different types of gears based on your requirements

Gears can be made from different materials such as plastic and metal, depending on their use. Different types of gears have different applications. Let's understand each of them by referring to the following table:

Gear Image	Gear Name	Advantages	Disadvantages	Practical Application
	Bevel	Quiet while in use and less friction	Complex manufacturing	Differential drive. Crown, conical, and double conical are different types of bevel gears.
	Spur	Simple design, high precision	Loud while in use	Clocks, pumps, and watering systems.

Gear Image	Gear Name	Advantages	Disadvantages	Practical Application
	Worm	Self-locking, high torque	Less efficiency and needs continuous lubrication.	Gear boxes and lifting mechanisms.
	Rack and Pinion	Simple design, high load-carrying capacity	More friction and limited movement	Steering systems, weighing scales, and others.

Table 5.1 - Different types of gears

In your LEGO BOOST kit, gears such as bevel, crown, conical, double conical, and spur gears are provided. We used spur gears to construct our robot.

Understanding some important terminologies

Before we learn more about gears and perform this chapter's activities, let's learn about some important terminologies:

- **Drive gear**: A gear that is connected to the motor.

- **Driven gear**: A gear that is connected to the drive gear.

- **Mesh**: This is when the drive and driven gears are connected to each other.

- **Gear ratio**: The ratio of *number of teeth of driving gear / number of teeth of driven gear.*

- **Torque**: The amount of force needed to bring your wheel into motion.

- **Gear up**: When your gear ratio is greater than 1, this is known as the gear up mechanism. With gear up, your driven gear moves at a faster speed than the drive gear and offers less torque.

- **Gear down**: When your gear ratio is less than 1, this is known as the gear down mechanism. With gear down, your driven gear moves at a slower speed than the drive gear and offers more torque.

Basically, whenever you need more power – for example, to lift a heavy load, go up a slope, and so on – you need high torque and less speed, which is what the gear down mechanism provides. When you are cruising on a highway at a high speed, you do not need a lot of power but a lot of speed. This is when the gear up mechanism should be used.

Let's calculate the gear ratio for the following gear combinations. The gear on the left-hand side in the first figure is the drive gear, while the one on the right-hand side is the driven gear:

Gear Combination	Number of Teeth in Drive Gear	Number of Teeth in Driven Gear	Gear Ratio
	40	8	40/8 = 5. Here, the gear ratio is greater than 1, which is why it is called the gear up mechanism.

Table 5.2 - Calculating gear ratio

Now, let's start building a geared robot.

Building the geared robot

In this section, we will be building the following robot:

Figure 5.1

To do this, perform the following steps:

1. Take your LEGO BOOST Hub and insert six working AAA batteries into it before you start constructing your robot.

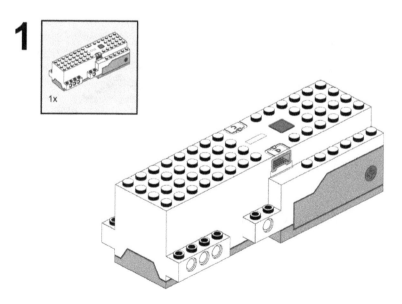

Figure 5.2

2.   Take two 4x6 bricks and connect them to the top and bottom of the BOOST Hub:

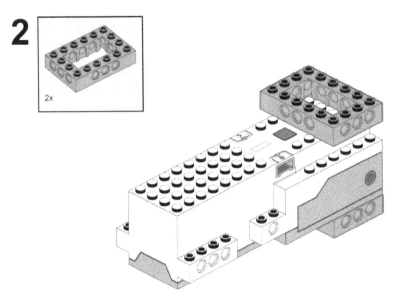

Figure 5.3

3.   Take two 2x5 plates and two 1x4 plates with knobs. Connect the 2x5 plates to the 4x6 brick. Then, connect the 1x4 plates with knobs to the back of the Hub, on both sides:

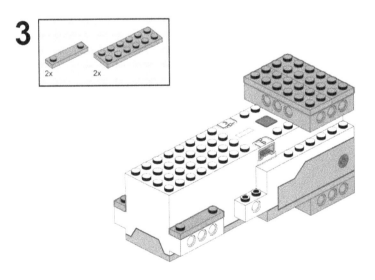

Figure 5.4

4. Take two 1x4 flat tiles and connect them to the 1x4 plates with knobs:

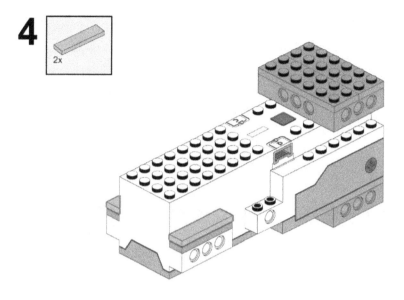

Figure 5.5

5. Take two 3M pegs and connect them to the last hole of the 4x6 brick, as shown here:

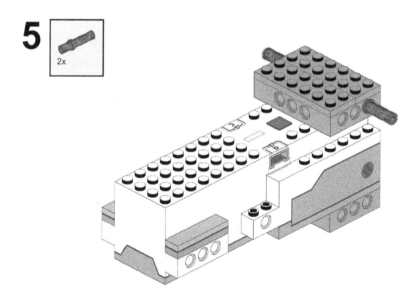

Figure 5.6

6.   Take two 1x2 bricks with holes. Connect them to the recently connected 3M pegs:

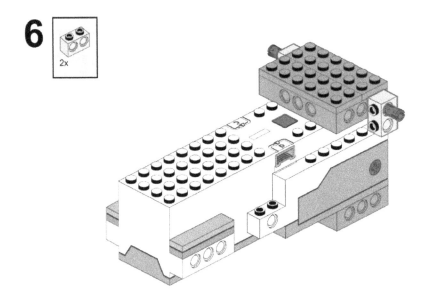

Figure 5.7

7.   Take two 4M cross axles and connect them to motor A and motor B.

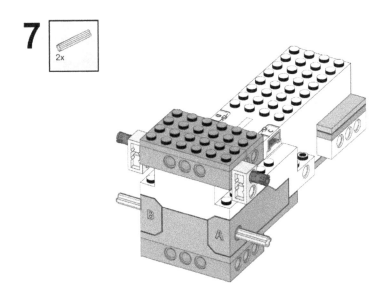

Figure 5.8

8. Take two eight-teeth spur gears and connect them to the 4M cross axles on motor A and motor B.

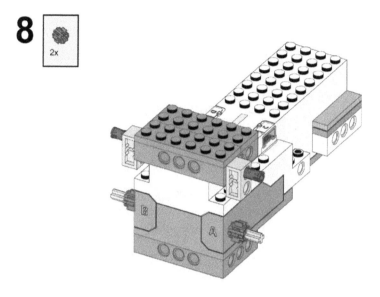

Figure 5.9

9. Take two 7M beams. Connect them to the 1x2 brick with holes and the 4M cross axles by their first and fifth holes, respectively:

Figure 5.10

10. Take two 24-teeth spur gears and place them beneath the eight-teeth spur gears:

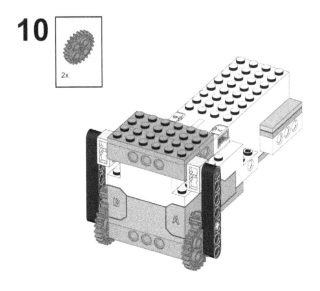

Figure 5.11

11. Take two full bushings, two 5M cross axles with end stops, two wide rims with crosses, and two normal wide tyres. Connect the rims to the tyres. Now, insert the stop axles through the full bushing as well as the wheels, and eventually pass it through the last hole of the 7M beam, the 24 teeth gears and the last hole of 4x6 plates as shown in the following figure:

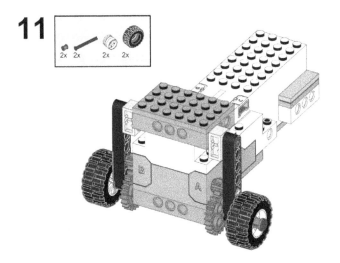

Figure 5.12

12. Take four black pegs and connect them to the BOOST Hub, beneath the orange flat tiles, as shown here:

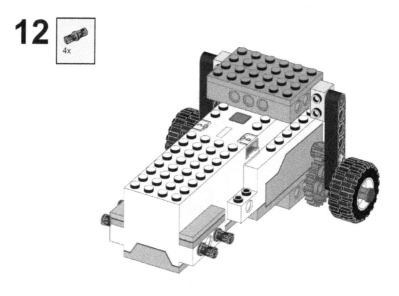

Figure 5.13

13. Now, take two 3x5 angular beams and connect them to these pegs on both sides:

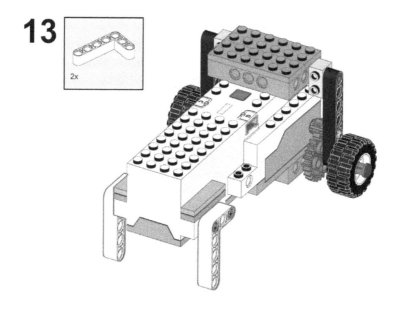

Figure 5.14

14. Now, take a 9M cross axle, two hub wheels, and four half bushings. Connect them so that they go through the angular beams, as shown in the following figure. This is your driven wheel:

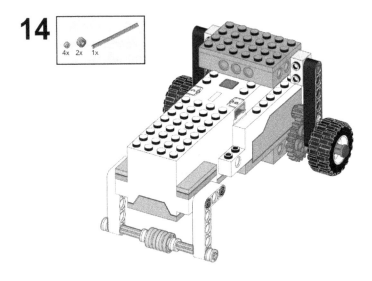

Figure 5.15

15. Now, take two 1x2 bricks and two 2x2 roof tiles. Connect them to the BOOST Hub, as shown in the following figure:

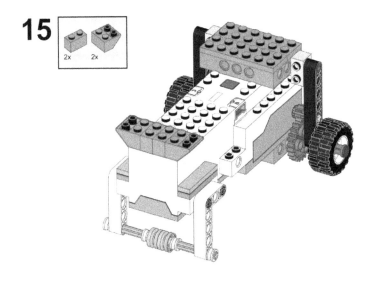

Figure 5.16

16. Take two 1x1x1 1/3rd white and blue bricks with arcs and the same blue bricks. Connect them to both sides of the roof tiles, as shown in the following figure:

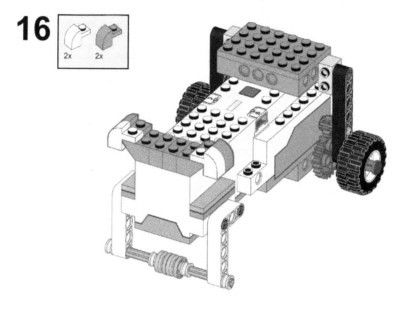

Figure 5.17

17. Take two 1x flat tiles and connect them to the bricks with arcs:

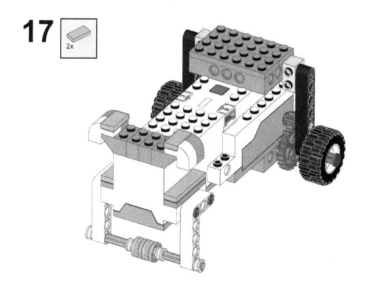

Figure 5.18

18. Take a 2x4 white brick with a face design and connect it to the LEGO bricks, as shown in the following figure. Now, take two round plates with knobs and place them on top of this white brick:

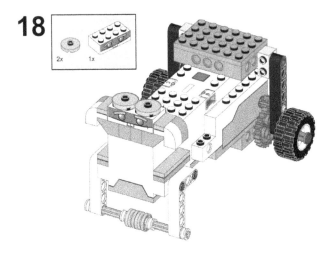

Figure 5.19

19. Take two wings with 4 mm shafts and connect them to the circular plates with knobs. This will add an aesthetic element to your gear robot:

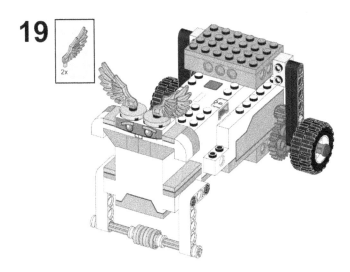

Figure 5.20

Great! Now that we've finished constructing the robot, we can start programming it so that it can perform tasks.

Let's code the robot in gear up and gear down mechanism

In this chapter, you will not be learning about any new programming blocks. Instead, you will be using the programming blocks you learned about previously in this book and carrying out some fun activities with your BOOST robot. Before you get started with these activities, you will need to set up the following:

- A clear path for your robot to move around – at least 3 feet long.
- Fix the starting point of the robot. Mark it with a clear demarcation. You will be starting your robot from this position in every activity.
- A measuring tape.

Now let's perform a few activities.

Activity 1

Program your robot to move forward for five rotations at 70% speed. You can change the brick's light if you wish. Measure the time taken, and distance traveled, and then fill in the following table:

Gear Ratio	Time Taken (seconds)	Distance Traveled (cm)

Table 5.3 - Measuring the time taken and distance traveled by the robot

The sample code for the activity is given here:

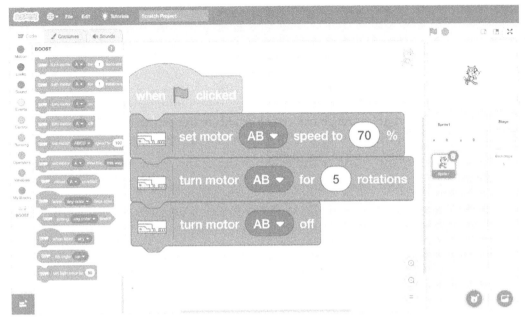

Figure 5.21 - Sample code

Now, swap your drive gear and driven gear.

When you need to swap the drive and driven gears, please refer to the following figure, as this will make things much easier for you:

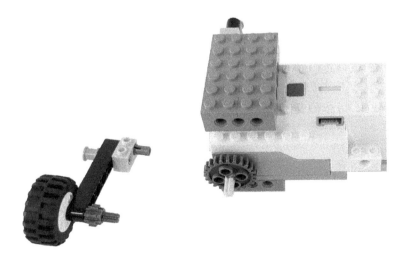

Figure 5.21 – Process of swapping the drive and driven gears

When you do this, your robot drive wheels will look like this:

Figure 5.22 – Gearing up the robot

Now, let's perform another activity.

Activity 2

Program your robot to move forward for five rotations at 70% speed. You can change the brick's light if you wish. Measure the time taken and distance traveled, and then fill in the following table:

Gear Ratio	Time Taken (seconds)	Distance Traveled (cm)

Table 5.4 - Measuring the time taken and distance traveled by the robot

Select the right options from the following statements:

The time it takes to finish the given number of motor rotations in gear up is (the same as/ more than/less than) in the gear down mechanism. The distance traveled for the same motor rotations in gear up is (more than/less than) the gear down mechanism.

Do you know that you can calculate the speed of your robot with a simple formula?

Speed = (Distance Traveled) / (Time Taken)

Can you use your math skills and calculate the speed of your robot in *Activity 1* and *Activity 2*? Use the spaces provided to do the calculations.

The speed of the robot in *Activity 1* = _____ cm/second.

The speed of the robot in *Activity 2* = _____ cm/second.

Time for a challenge

Challenge #1

Get your robot back in the gear up combination. Now, create a ramp using the resources available in your home. Try implementing a different height for this ramp and reach a point where your robot is no longer capable of going up that ramp in this gear up combination!

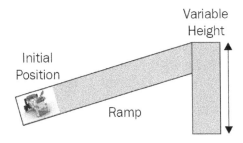

Figure 5.23 – Representation of the setup

The height of this ramp = _____ cm.

Now, change your gear combination, so that it's in the gear down combination and see whether it can cross this ramp. I am sure it will. This is because the robot is slower, but its torque is quite high! Can you find a height for this ramp where this robot can't climb, even when using the gear down combination?

The height of this ramp = _____ cm.

One major conclusion we can draw from this chapter is that when you try to gain one thing, you lose the other. For example, when you gain speed, you lose power and vice versa! It takes a lot of engineering prowess to achieve high-speed, high-torque mechanisms. Later in this book you will learn more about changing the axis of force using gears.

Challenge #2

Do you remember building and coding a tabletop fan? Can you make use of gears and rebuild this fan so that it moves much quicker to spread cool air around the room?

Summary

In this chapter, you learned about how to use gears. You also learned about the different types of gears that are available in the market, along with their advantages, disadvantages, and practical applications. To understand the difference between the gear up and gear down mechanisms, you performed two activities and compared some critical parameters such as time taken and distance traveled for a specific number of motor rotations. To understand the importance of high torque in certain conditions, such as going up a ramp, you tried to crack the challenges provided and understood the impact of the gear down mechanism in achieving high torque.

In the next chapter, you will be building an amazing forklift that will be able to lift the load from one place to another, all by itself. A forklift is widely used in warehouses and industries to make human work easy. The forklift will be made with the application of simple machines and a gear mechanism.

Further reading

You can learn more about gears at `https://www.thomasnet.com/articles/machinery-tools-supplies/understanding-gears/`.

Gears can have a lot of wear and tear issues, but if maintenance is done at the right time, these gear mechanisms can add a lot of power and speed to your machinery, while also helping change the direction of the force you're applying.

6
Building a Forklift

Have you ever seen a machine being used to move heavy loads from one place to another in industries, warehouses, or even at malls? Have you ever wondered what this machine is called? Such a machine is called a forklift! Now, you might be wondering why it's called a forklift. This is because it lifts heavy loads on its two fork-like tongs at the front and places them where they need to be placed. Forklifts are compact and used in various places, indoor as well as outdoor. In this chapter, you will be building your own forklift using the BOOST Hub and LEGO pieces and coding it to pick up loads and place them elsewhere. This chapter will help you gain insights into how forklifts work and help you build and code one using your LEGO BOOST kit:

Figure 6.1

In this chapter, we will cover the following topics:

- Building the forklift robot
- Let's code the robot to lift different loads
- Time for a challenge

You will be using the gear down/power up mechanism to lift the heavy loads with your forklift. Two hub motors will act as the drive base for the forklift, and the external motor will be used to operate the forklift mechanism. You will be building two separate things – a forklift and a load base. Let's get started!

Technical requirements

In this chapter, you will need the following:

- A LEGO BOOST kit with 6 AAA batteries, fully charged
- A laptop/desktop with the Scratch 3.0 programming language installed and an active internet connection
- A diary/notebook, along with a pencil and eraser

Building the forklift robot

In this chapter, we will be building the robot shown in the following figure:

Figure 6.2

Follow these steps to build the forklift robot:

1. Take your BOOST Hub and ensure that the batteries are fully charged:

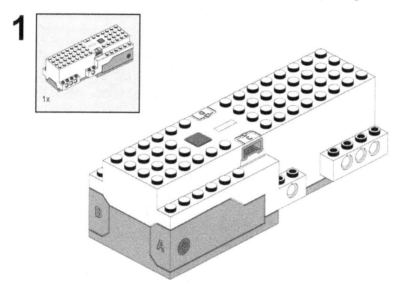

Figure 6.3

2. Take one 2x4 studded plate and one 2x4 white brick with a face design. Attach them to the BOOST Hub:

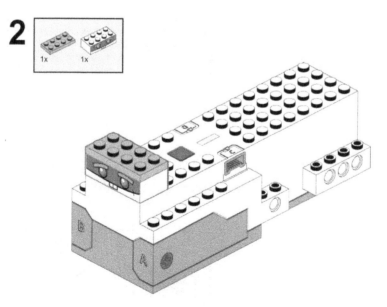

Figure 6.4

3. Take the LEGO bricks indicated in the following figure and stack them up on the recently connected LEGO bricks on the BOOST Hub:

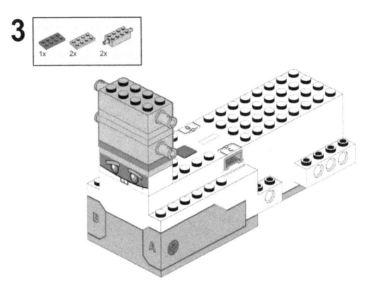

Figure 6.5

4. Take two 16M Technic beams and connect them to the stack of LEGO bricks you just placed, as shown here:

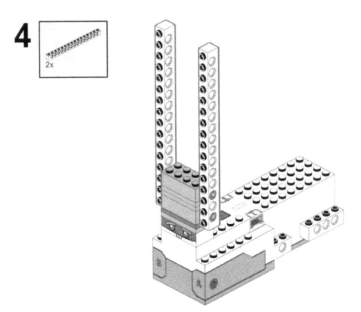

Figure 6.6

5.   Take two black 2M pegs and connect them to the Technic beams:

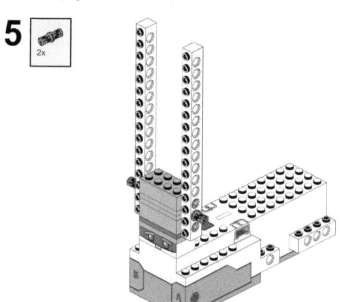

Figure 6.7

6.   Take two 10M Technic beams and attach them to the pegs that you just connected:

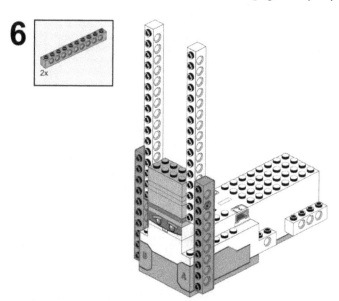

Figure 6.8

7. Take two 2x4 plates and connect them to the 16M and 10M Technic beams to make the following 10M Technic beam connection rigid:

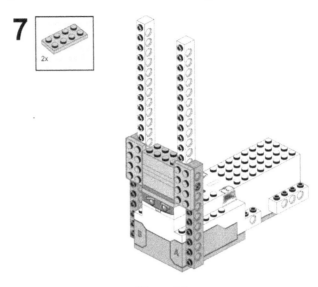

Figure 6.9

8. Take the full bushing, 5M cross axle with stops, and the wheels, as shown in the following figure. Now, insert the wheel at the top of the axle and insert the full bushing. Connect this attachment to the motor's A and B axle slots through the 10M Technic beam:

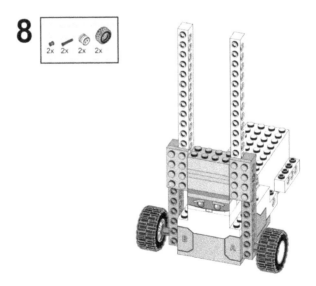

Figure 6.10

9. Take one connector bush with friction and a cross axle and connect it to the Technic beam, as shown here:

Figure 6.11

10. Take one 7M beam and one 2M cross axle extension. Connect them to the recently connected bush:

Figure 6.12

11. Take one connector bush with friction plus a cross axle and a 7M beam. Connect them to the 16M Technic beam on the other side:

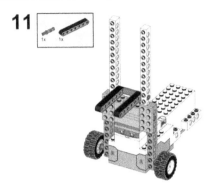

Figure 6.13

12. Now, take four 1x6 plates. Connect these two plates. Make two such sets and connect them to the top of the 16M Technic beam:

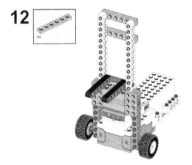

Figure 6.14

13. Take one 2x6 plate and connect it to the 16M Technic beam so that it's in-between the two 1x6 orange plates:

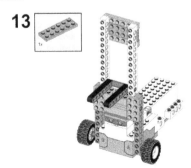

Figure 6.15

14. Take the external motor from your BOOST kit and connect it to the LEGO plate base that you just built. All the studs of your motor will be connected to the plates:

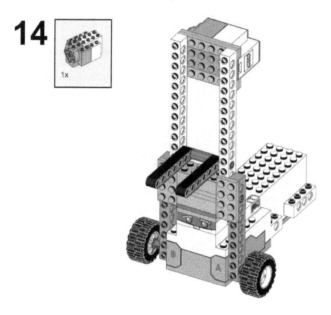

Figure 6.16

15. Take one 8M cross axle with stop, two double cross blocks, and two 9M beams. Connect them to the 16M Technic beam, as shown here:

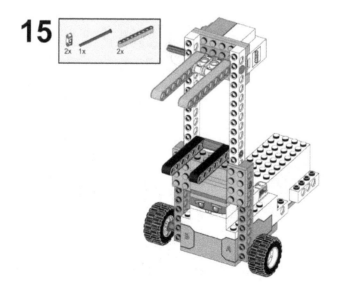

Figure 6.17

16. Now, take one 4M cross axle and insert it through the double-cross blocks and 9M beam. It should be inserted through the third hole from the motor's side:

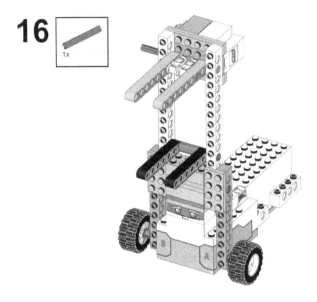

Figure 6.18

17. Take one full bush, one 3M cross axle with end stop, and one 8-teeth bevel gear. Insert the half bush toward the end stop side of the axle and then insert the gear. Now, attach this connection to the axle hub of the motor. This will act as your drive gear:

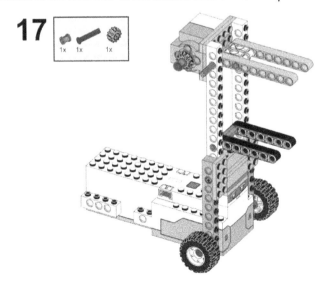

Figure 6.19

18. Now, take one 40-teeth bevel gear and one full bush. Connect this gear to the other axle and insert the bush at the top:

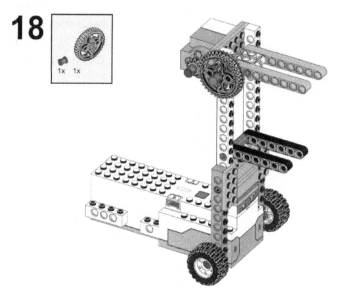

Figure 6.20

19. Take two 16M Technic beams and place them at the third hole of the 9M orange beam:

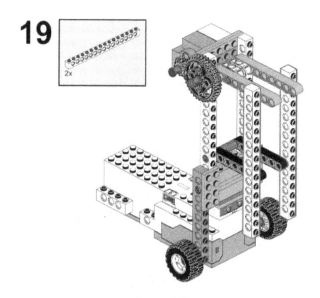

Figure 6.21

20. Now, take two 8M cross axles with stops and two tubes with double holes. Place the tubes between the two orange beams and pass the axel through them, ensuring that both 16M Technic beams are connected. Make sure that the open end of this axle comes out on the drive gear's side. Similarly, connect the 7M black beams to the 16M Technic beams through this axle and tube, as shown here. This will be connected to the sixth hole from the bottom of the 16M Technic beam:

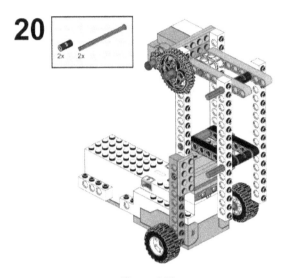

Figure 6.22

21. Take two half bushings and connect them to both axles:

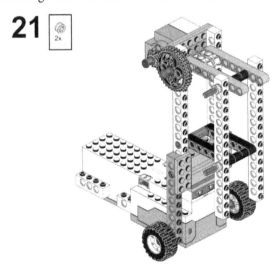

Figure 6.23

22. Take two 3x5 angular beams and four friction snaps with cross holes. Connect them to the bottom of the 16M Technic beam on both sides, as shown here:

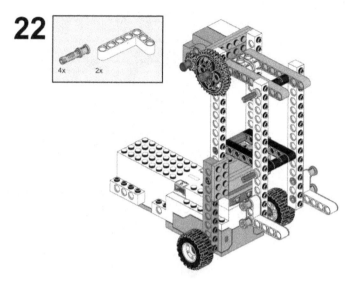

Figure 6.24

23. Take two steering knuckle arms and connect them to the 3x5 angular beams:

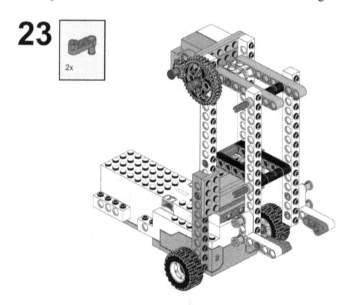

Figure 6.25

24. Take two catches with cross holes and connect them to the knuckles, as shown here:

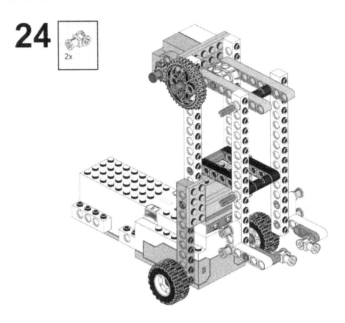

Figure 6.26

25. Take two half bushings and one full bushing. Arrange them so that they're in the pattern shown in the following figure, near the knuckles:

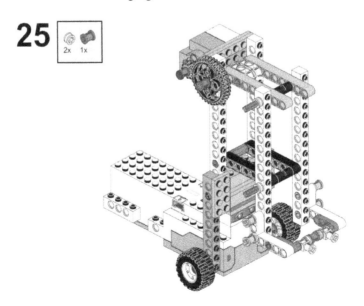

Figure 6.27

26. Take one 9M cross axle and pass it through the knuckle arm from one end, as well as the last hole of the angular beam. Ensure that all three bushings are connected via the same pattern that you formed in the previous step:

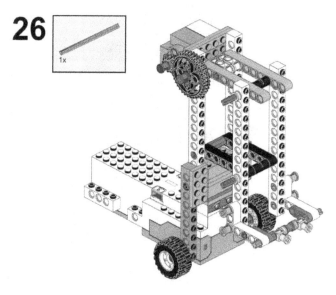

Figure 6.28

27. Take two 7M cross axles and two 2M dampers. Connect the axles to the catch and dampers at the other end of the axle:

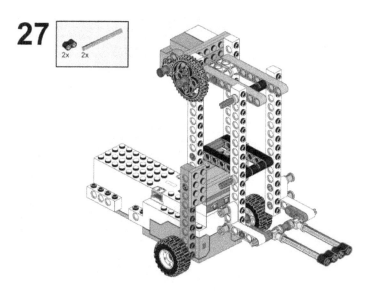

Figure 6.29

28. Now, let's build the rear/driven wheels. Take two 2M cross axles with snaps and friction. Connect them to the last hole of the rear end of your BOOST Hub:

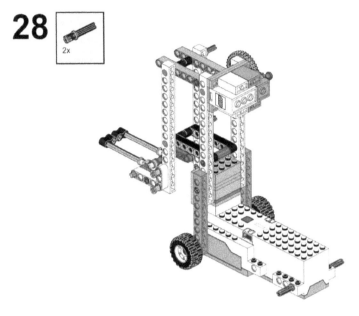

Figure 6.30

29. Take two half bushings and connect them to these 2M cross axles:

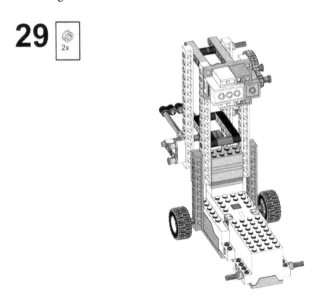

Figure 6.31

30. Take four 4x4 round plates with snaps and connect to make two pairs. Then, attach them to the axles on both sides:

Figure 6.32

31. Now, take two half bushings and connect them to the wheels to keep them at the same position on both sides:

Figure 6.33

32. Now, take one 2x4 brick and one 2x2 brick. Connect them to the external motor:

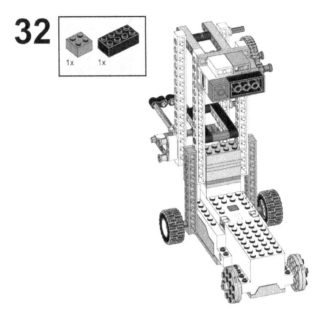

Figure 6.34

33. Take two 2x2 bricks with snaps and crosses. Connect them to the recently attached LEGO bricks, as shown here:

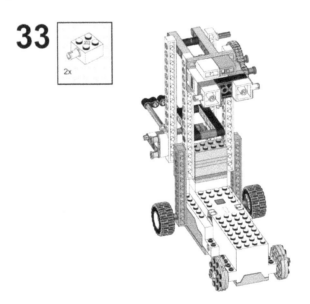

Figure 6.35

34. Take two 2x6 plates and connect them. Now, attach this to the 2x2 bricks with snaps and crosses:

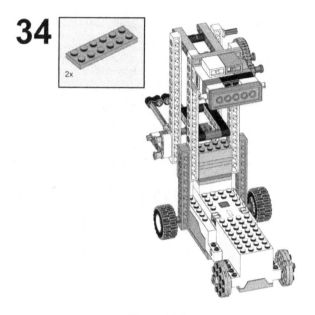

Figure 6.36

35. Take two 1x2 bricks with crosses and connect them to the 2x6 plates, as shown here:

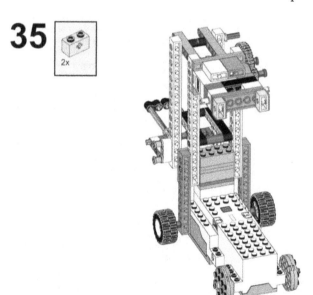

Figure 6.37

36. Take two 2M cross axles. Connect them to the 1x2 bricks with crosses:

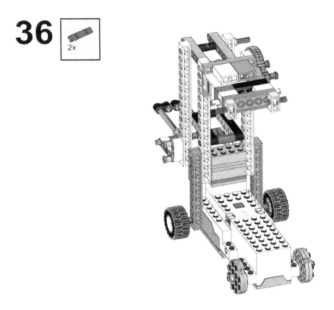

Figure 6.38

37. Take two 3x5 angular beams and connect their first and third holes to the 2x2 brick with snap and the 2M axle with the 1x2 brick with a hole on both sides, respectively:

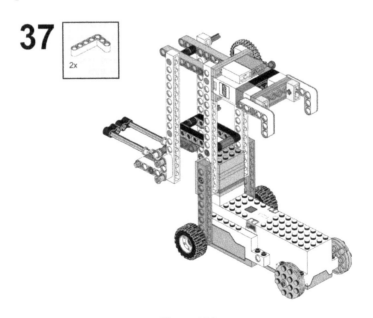

Figure 6.39

38. Take four black pegs and connect them to the 3x5 angular beams, as shown here:

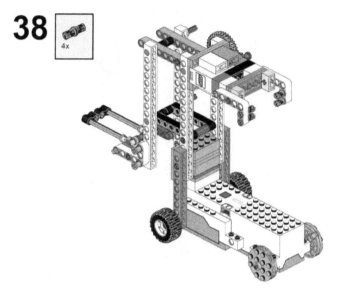

Figure 6.40

39. Now, take two black pegs and two 8M Technic beams. Connect both pegs to the 8M Technic beams at the last hole. Then, connect this 8M Technic beam to the angular beam, as shown in the following figure:

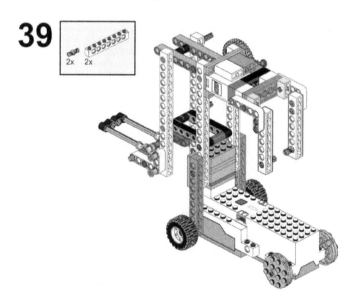

Figure 6.41

40. Take one 4x6 brick and connect it to the Technic beams, as shown here:

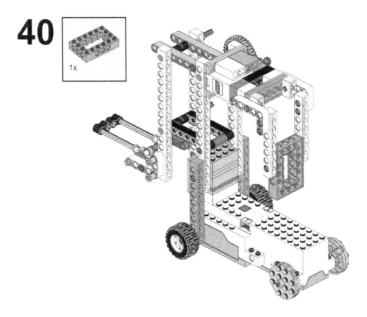

Figure 6.42

41. Now, take two connector bushes with cross axles. Connect them to the last holes of the 4x6 brick, as shown here:

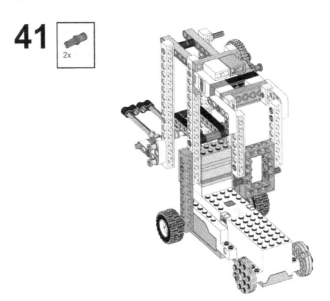

Figure 6.43

42. Take two more connector bushes with cross axles and two design shape tubes with cross holes. Connect them to the 4x6 brick, as shown in the following figure:

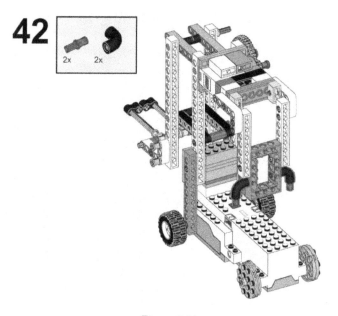

Figure 6.44

43. Take two more 4x6 bricks and connect them to the design tubes on both sides:

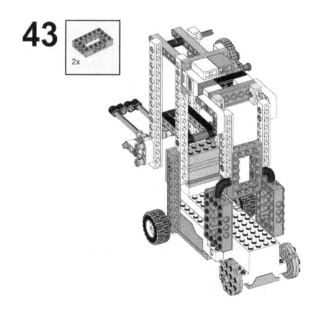

Figure 6.45

44. Take two 2x4 bricks. Connect them to the inside of the 4x6 bricks:

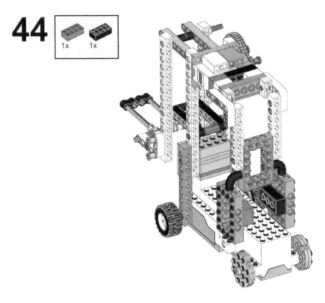

Figure 6.46

45. Take two 2x4 plates and one 2x4 brick. Connect the orange plate to the red brick. Stack the black plate on top of the black brick and connect this to the other internal side, as shown here:

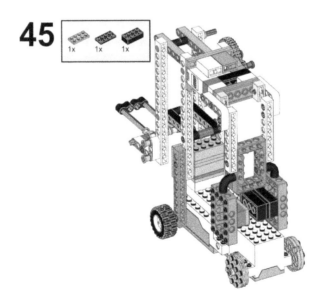

Figure 6.47

46. Now, take two 2x2 bricks and connect them to the 2x4 black brick, as shown here. This will add strength to the two 4x6 bricks' connection:

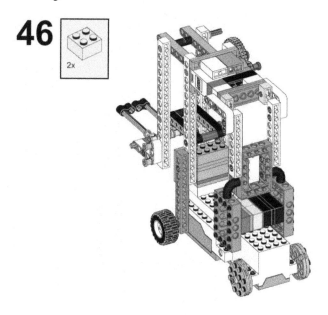

Figure 6.48

Your forklift is now ready!

Now, let's build the stand that we will place the load on so that your forklift can lift it:

1. Take four 2x2 round bricks with holes on their sides:

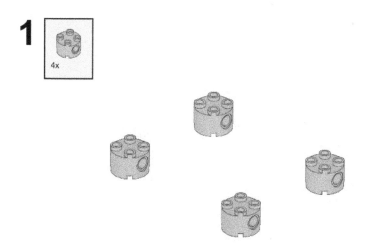

Figure 6.49

2. Take four red 1x2 bricks with cross holes. Connect them to the 2x2 round bricks, as shown in the following figure:

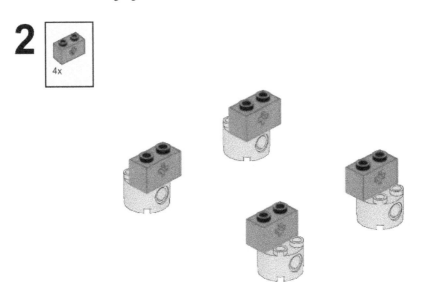

Figure 6.50

3. Now, take four green 1x2 bricks with crosses and stack them on top of the recently attached red 1x2 bricks with crosses:

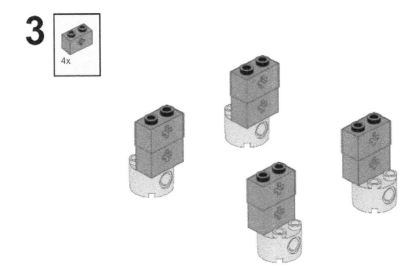

Figure 6.51

4. Now, take four blue 1x2 bricks and stack them on top of the green bricks:

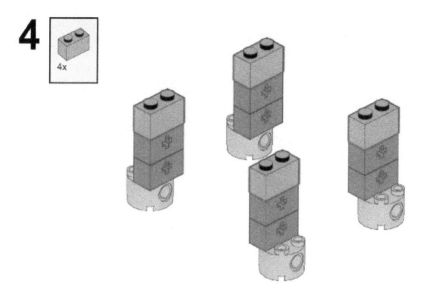

Figure 6.52

5. Now, take two 1x8 plates and connect them to the blue bricks, as shown in the following figure:

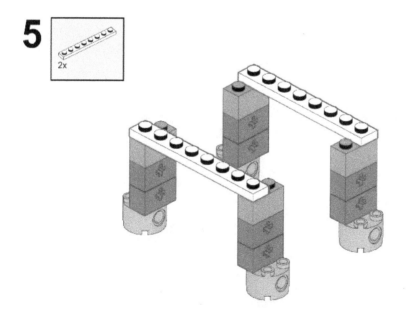

Figure 6.53

6. Now, take two 2x8 plates and connect them to the blue bricks, besides the 1x8 plates:

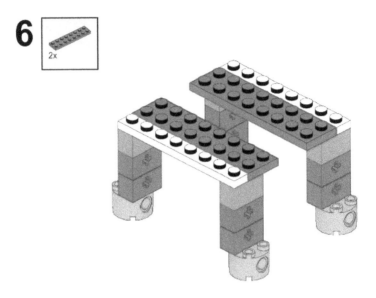

Figure 6.54

7. Now, take two 4x4 plates and one 2x8 plate. Place the 2x8 plate in-between the other 2x8 plates and connect them using the 4x4 plates:

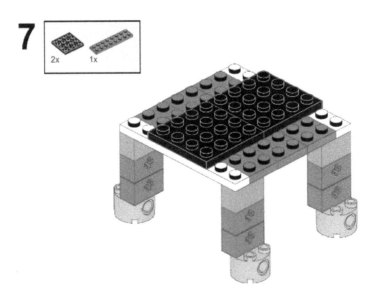

Figure 6.55

8. Take four 4x4 plates with 1/4th circles and connect them to the other plates to make a platform, as shown here:

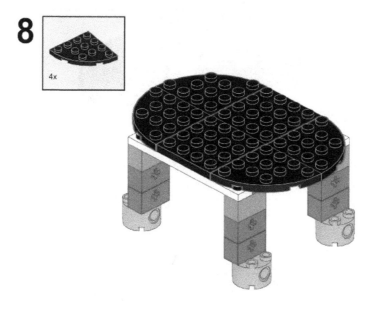

Figure 6.56

9. Take two 1x4 plates. Flip the base and connect these plates, like so:

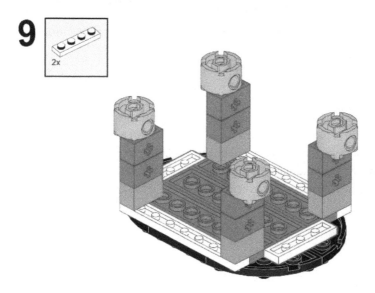

Figure 6.57

10. Now, flip it back over. With that, your base is ready!

Figure 6.58

Connect the external motor to port C. Since the forklift and the base are ready, let's complete the necessary programming tasks.

Let's code the robot to lift different loads

Let's start coding this forklift. You will be using all three of the robot's motors for the first time, which is going to be quite exciting.

Activity #1

Mark the initial position of your load base, as well as the forklift robot. Ensure there's at least 30 cm distance between them and try to keep them in a straight line. Now, program your robot to move forward until it reaches the load. The robot must stop so that the forks go underneath the load base. Try to use rotations-based programming for the drive motors for accuracy. Please use the given space to write the algorithm.

Now that you are there, you need to pick up this load. To do so, you will have to move the load motor so that it is connected to port C. As you may recall, you can program your motor so that it moves for a few seconds. We discussed this in *Chapter 2, Building Your First BOOST Robot – Tabletop Fan*, and *Chapter 3, Moving Forward/Backward Without Wheels*. Now, it is time to use this feature. Whenever you are programming a grabber (a forklift, in this case), try to code it using seconds-based programming so that the forklift is never stuck in one place. After some trial and error, you will find out how many seconds your motor takes to lift the load at a specific speed. Keep this speed and time constant and note it down in your book. Also, find out what the parameters are for putting the load down with the grabber. Write these values in the following table:

Action	Motor Speed/Power	Time Taken
Forklift Up		
Forklift Down		

Table 6.1 - Measuring motor speed and time taken

The sample code for the activity is given here:

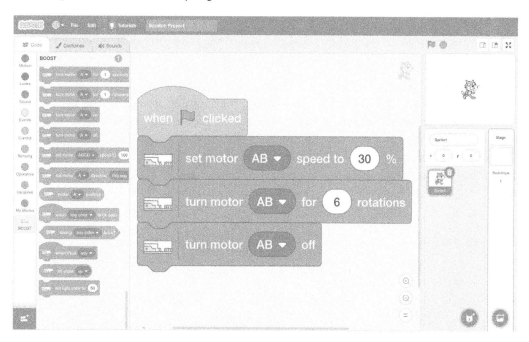

Figure 6.59 - Sample code

Let's move on to the next activity.

Activity #2

Program your robot to lift the forklift's forks for three rotations at a power of 50 and play a sound file of your choice. Observe this for 2 minutes to understand what happens with the grabber/forklift motor.

Activity #3

Program your robot to lift the forklift's forks for 4 seconds at a power of 50 and play a sound file of your choice. Observe what happens!

In which of the above activities, was the sound played? Activity #2 or #3? Select the correct option. Do you know why? In our rotations-based programming activity, the number of rotations was far more than what was needed, and the motor could not finish those rotations. Hence, the next block was not executed, and your robot got stuck. However, in seconds-based programming, even though the time was much more than what was needed, the block was executed for 4 seconds, and then the sound file was executed. Hence, we should use seconds-based programming in such applications.

The sample code for the activity is given here:

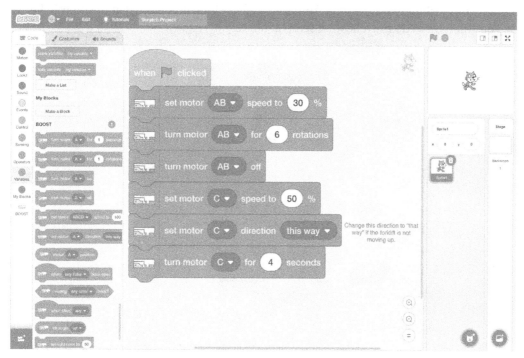

Figure 6.60 - Sample code

Let's look at the next activity.

Activity #4

Code your forklift to lift this load base and return to the starting position of the robot and place the load base there. Please use the space to write the algorithm, before you start coding.

Now, place the forklift robot and the load base at their initial positions.

Activity #5

Program your forklift to lift the load from its initial position and place it at position B. For this, you will have to do the following setup at home:

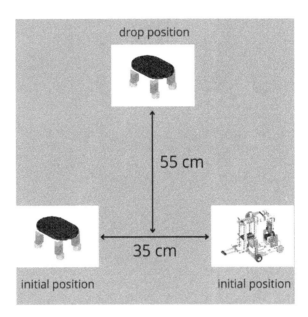

Figure 6.59

You will have to go through some trial and error to find out the number of rotations for the following:

1. Turning the robot right = _____ rotations.

2. The number of rotations to cover 35 cm = _____ rotations.

3. The number of rotations to cover 55 cm = _____ rotations.

Great! You can try running the robot on the same route with different types of loads placed on the load base, such as a book, a pencil holder, or a lunch box.

Time for a challenge

Challenge #1

Again, place your forklift robot and the load base at their initial positions. Place a small book of your choice on this load. Also, keep one TV/AC remote or a compass handy. Now, program your robot to do the following:

1. Pick up the load base with the book from its initial position and drop it at point A.

2. Place the remote control of your TV/AC or your compass on the load base and drop it at point B.

3. Now, program your robot to return to its base.

You will need the following setup for this challenge:

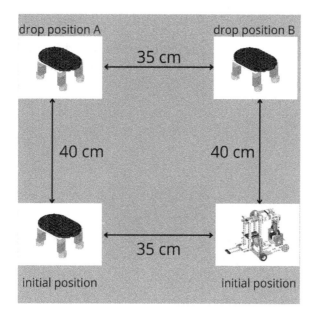

Figure 6.60

Let's try another interesting challenge.

Challenge #2

Can you find out what is the maximum load that your forklift can pick up? _____ kg.

Did you know that you can only pick up and drop stable loads such as boxes with this forklift? Try lifting a water bottle with this forklift and place it somewhere else. Your bottle will fall. We will look at this in more detail in the next chapter.

Summary

In this chapter, you learned about the practical application of the forklift mechanism. You learned how to make all three motors move when connected to the BOOST Hub. In this chapter, you executed seconds-based programming for your forklift and understood the advantage of this feature.

In the next chapter, you will build a helicopter with moving blades and solve various challenges. Again, you will be applying your learning of gears to build this helicopter. There will be super fun activities where you will rescue and deliver items to a remote location using this helicopter.

Further reading

You can learn more about the application of a forklift at `https://www.youtube.com/watch?v=Ne2s2AtUUUU&ab_channel=ToyotaForklift`.

You can increase the height of loading and unloading your forklift with the help of various mechanisms, such as chains, pulley systems, and so on. You will learn about this in more detail in the coming years when you further your robotics courses/exploration.

7
Building a Helicopter

We are always fascinated by the way we see airplanes and helicopters flying. They are the fastest form of transport today. While planes are used for long-haul as well as short-haul routes, helicopters are quite handy for traveling within the city as well! While airplanes need a long runway to take flight, helicopters can fly without a runway. Just like fighter planes, helicopters are also used in warfare. To achieve flight, various factors must be taken into consideration, such as thrust, drag, weight, and lift, all of which need to be balanced. Although we cannot build a flying helicopter using our LEGO bricks and BOOST Hub, we will build a prototype helicopter in this chapter and go through some fun activities. This helicopter will have rotating blades on the top as well as at the back, as shown in the following figure:

Figure 7.1 – Helicopter

In this chapter, we will cover the following topics:

- Building a helicopter
- Let's code the robot to perform various tasks
- Time for a challenge

Technical requirements

In this chapter, you will need the following:

- A LEGO BOOST kit with 6 AAA batteries, fully charged
- A laptop/desktop with the Scratch 3.0 programming language installed and an active internet connection
- A diary/notebook, along with a pencil and eraser
- Sticky notes

Building a helicopter

In this chapter, we will be building the following helicopter:

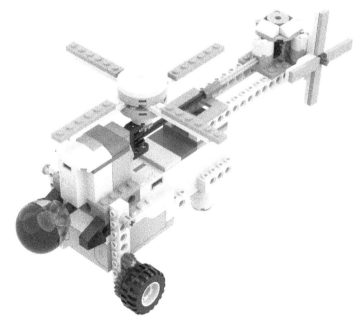

Figure 7.2

Follow these steps to build the helicopter:

1. Take your BOOST Hub and ensure that its batteries are fully charged.

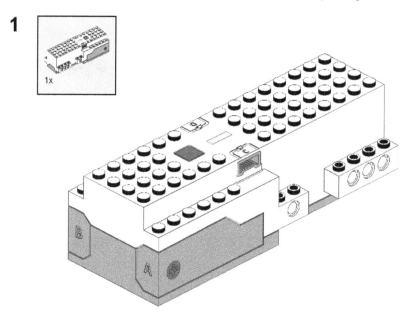

Figure 7.3

2. Take two wing sections and connect them to the BOOST Hub.

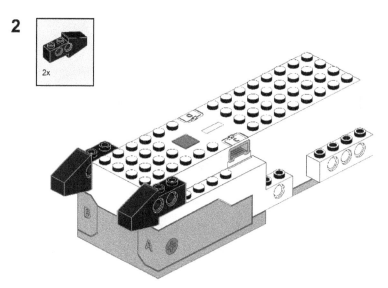

Figure 7.4

3. Take two 2x4 plates and one 4x6 brick. Connect them to the BOOST Hub, as shown in the following figure:

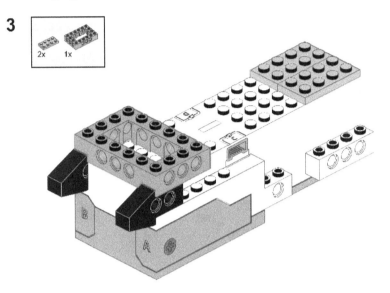

Figure 7.5

4. Now, take one blue flat tile, one yellow flat tile, two 2x4 red plates, and two 2x4 orange plates. Connect the blue and yellow tiles to the BOOST Hub, connect the 2x4 red plate to the recently attached 2x4 orange plate, and then connect the 2x4 orange plate to the 4x6 brick.

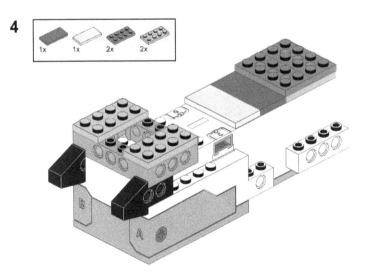

Figure 7.6

5. Take one 3M axle and connect it to motor A of the BOOST Hub.

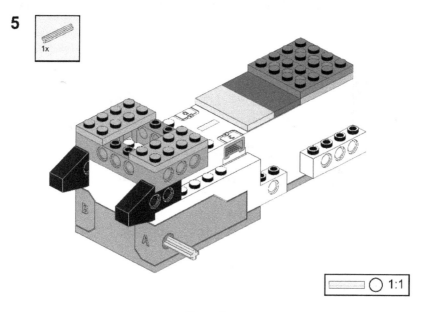

Figure 7.7

6. Take one 3M stopper axle, one connector peg, and one 1x8 Technic brick. Connect the 3M stopper axle to the 1x8 Technic brick and then attach the Technic brick to the 3M axle with the connector peg.

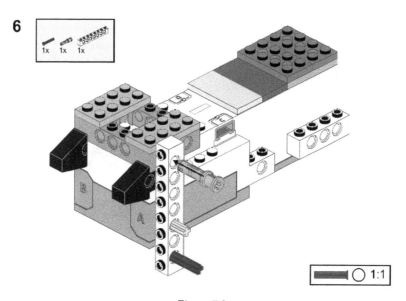

Figure 7.8

7. Take one 8-toothed spur gear and one 24-toothed spur gear. Connect the 8-toothed spur gear to the 3M axle and the 24-toothed spur gear to the 3M stopper axle.

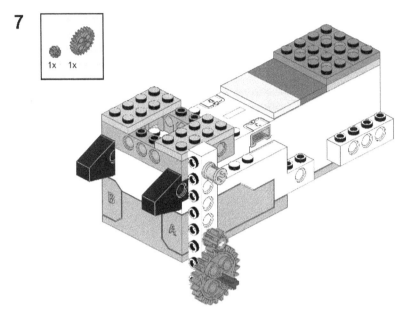

Figure 7.9

8. Take one 3M axle and connect it to motor B of the BOOST Hub.

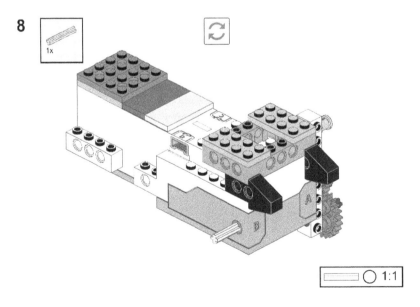

Figure 7.10

9.   Take one 3M stopper axle, one connector peg, and one 1x8 Technic brick. Connect
     the 3M stopper axle to the 1x8 Technic brick and then attach the Technic brick to
     the 3M axle with the connector peg.

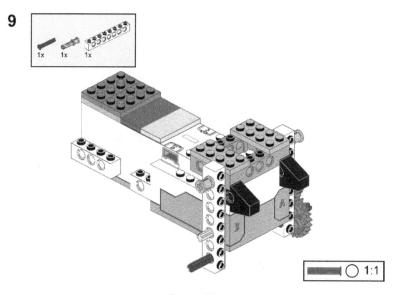

Figure 7.11

10. Take one 8-toothed spur gear and one 24-toothed spur gear. Connect the 8-toothed
     spur gear to the 3M axle and the 24-toothed spur gear to the 3M stopper axle.

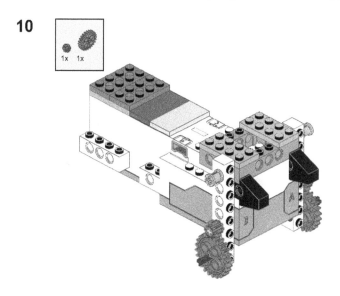

Figure 7.12

11. Take two rims and two tires and assemble them. Connect them to both sides of the recently attached 3M stopper axle.

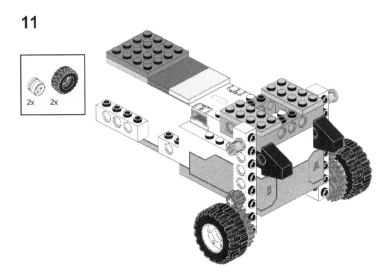

Figure 7.13

12. Now, take one 2x2 slide shoe round and three 2x2 round bricks. Connect them underneath the BOOST Hub.

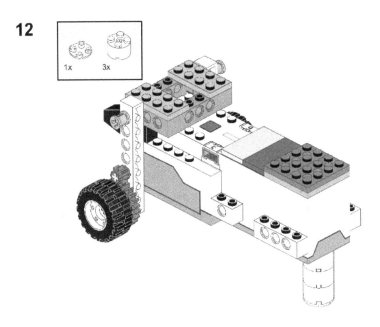

Figure 7.14

13. Take one external motor and two pegs with axles. Connect the pegs with axles to the motor.

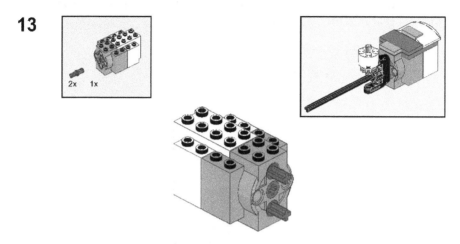

Figure 7.15

14. Take two 1x4 plates, one 3M beam with a fork, and one 12M axle. Insert the 12M axle into the motor and connect the 1x4 plates to the motor. Then, connect the 3M beam with a fork to the external motor.

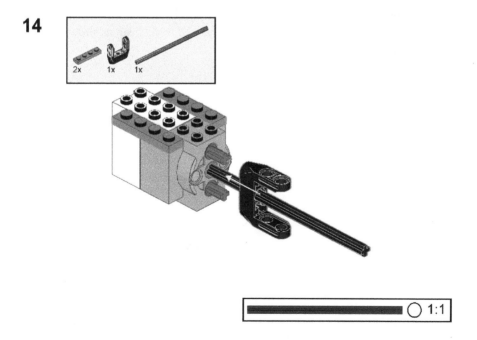

Figure 7.16

15. Take two 1x2 flat tiles, one 12-toothed bevel gear, one 2x4 flat tile, and one 3x4x2/3 plate with a bow. Connect the 3x4x2/3 plate with a bow to the motor. Then, connect the 2x4 flat tiles to the 3x4x2/3 plate, as shown in the following figure, and insert the 12-toothed gear into the 12M axle:

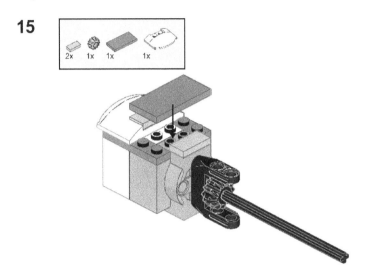

Figure 7.17

16. Now, take one half bevel gear and one 3M axle. Mesh the half bevel gear with the 12-toothed gear and insert the 3M axle into the half bevel gear, as shown in the following figure:

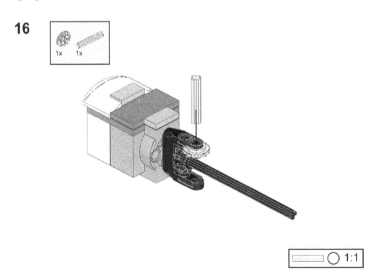

Figure 7.18

17. Take a 2x2 round brick. Connect it to the recently attached 3M axle.

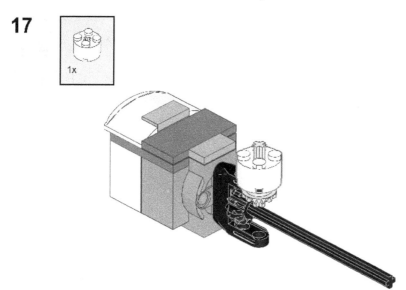

Figure 7.19

18. Now, connect the motor to the 4x6 brick, as shown in the following figure:

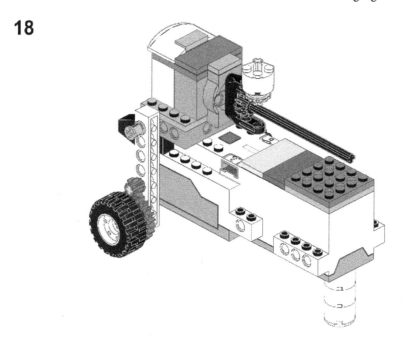

Figure 7.20

19. Now, take two 4x4 round plates and connect them, as shown in the following figure:

19

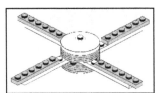

Figure 7.21

20. Now, take four 1x8 plates and connect them to the 4x4 round plate:

20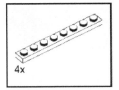

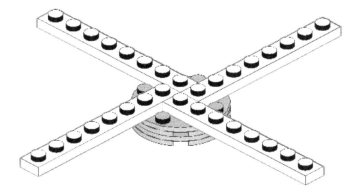

Figure 7.22

21. Take two 4x4 round plates and connect them to the recently attached 1x8 plate, like so.

21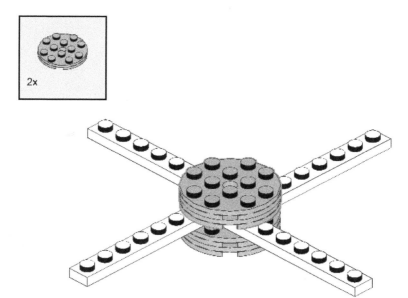

Figure 7.23

22. Take four 1x6 plates and connect them to the 1x8 plate.

22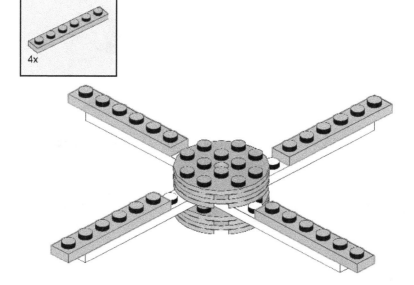

Figure 7.24

23. Take a round plate and connect it to the 4x4 round plate.

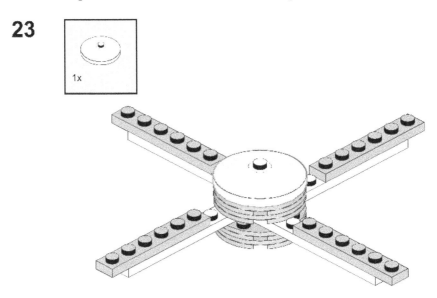

Figure 7.25

24. Now, connect the 2x2 round brick, along with the 3M axle, like so.

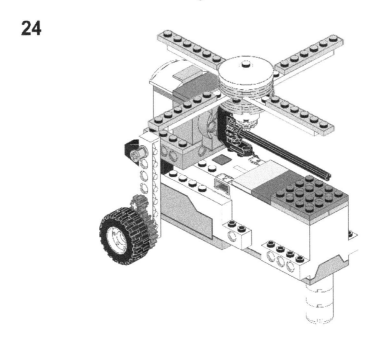

Figure 7.26

25. Take two 1x16 Technic beams and connect them to the 2x4 red plate.

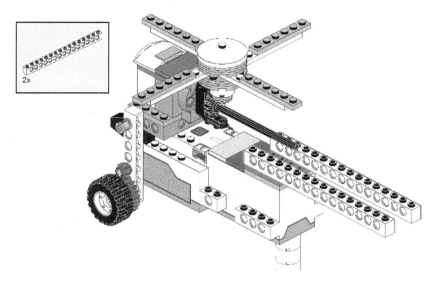

Figure 7.27

26. Take two 1x2 flat tiles and two 1x4 flat tiles. Connect them to the 1x16 Technic beams, as shown in the following figure:

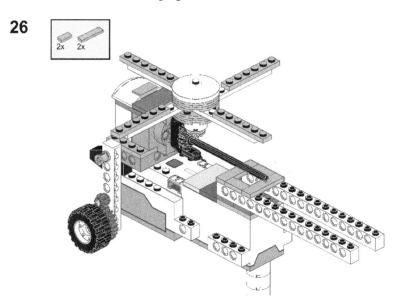

Figure 7.28

27. Take two 1x2 plates and two 2x4 plates. Connect them to the 1x16 Technic beams, as shown in the following figure:

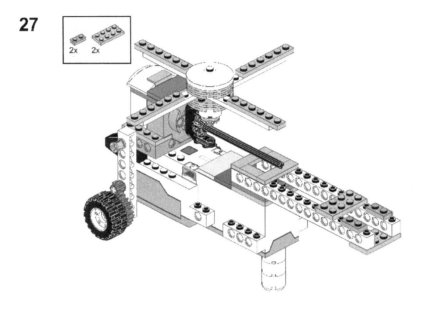

Figure 7.29

28. Take one axle extension and one 9M axle. Connect the axle extension to the 12M axle and insert the 9M axle into it.

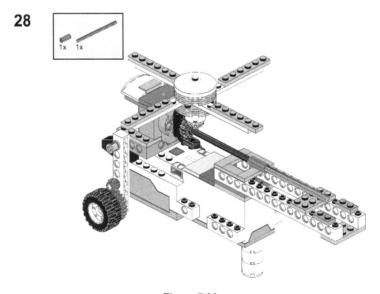

Figure 7.30

29. Take one 2x2 round brick with a hole in its side and one 2x2 round brick. Connect both bricks to the 2x4 plate with the 9M axle, as shown in the following figure:

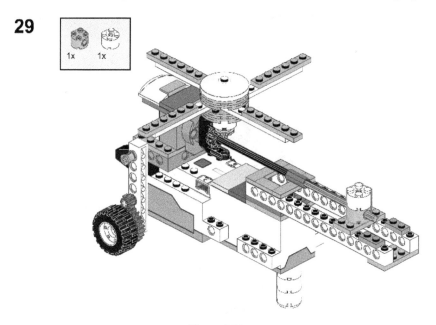

Figure 7.31

30. Take one 12-toothed bevel gear and connect it to the 9M axle.

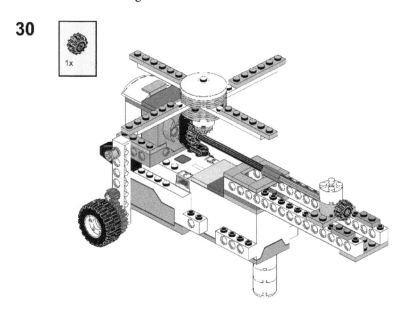

Figure 7.32

31. Take two 2x2 round bricks with holes in their sides and two 2x2 round bricks. Connect them to the 1x2 plate, as shown in the following figure:

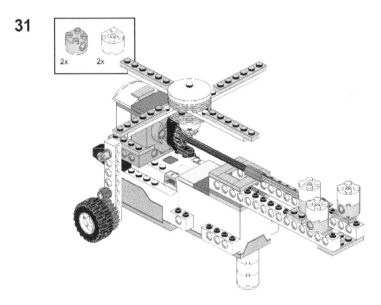

Figure 7.33

32. Now, take one half bevel gear and one 8M stopper axle. Pass the 8M stopper axle through the 2x2 round brick with a hole in its side and connect the half bevel gear to it, as shown in the following figure:

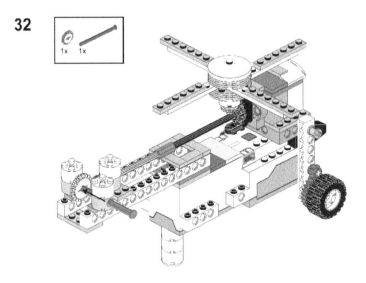

Figure 7.34

33. Take one 2x2 round plate and one 4x4 round plate. Connect the 2x2 round plate to the middle of the 4x4 round plate, like so.

33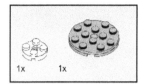

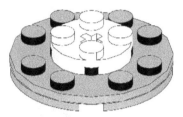

Figure 7.35

34. Now, take four 1x2x2/3 roof tiles. Place them so that there's one on each side of the 4x4 round plate.

34

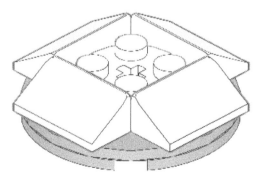

Figure 7.36

35. Take a 2x2 flat round tile and connect it to the 2x2 round plate.

35

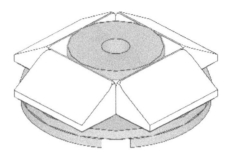

Figure 7.37

36. Now, connect this part to all three 2x2 bricks, as shown in the following figure:

36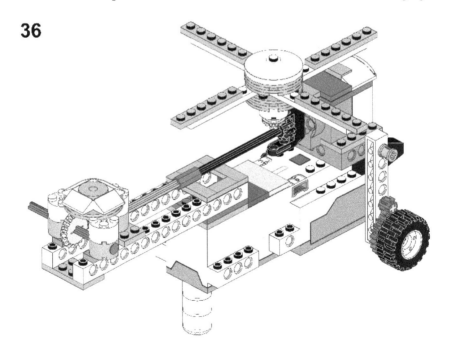

Figure 7.38

37. Take two Technic rotor blades and connect both so that they fit together, as shown in the following figure:

37

Figure 7.39

38. Take four 1x4 plates and connect them to the recently attached rotor blades.

38

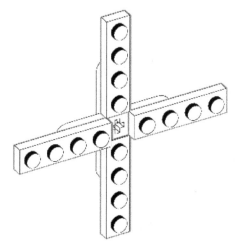

Figure 7.40

39. Now, take four 1x4 plates and place them on top of the 1x4 plates.

39

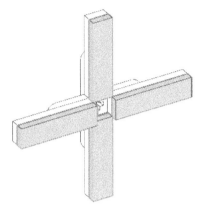

Figure 7.41

40. Now, connect this part to the 8M stopper axle, as shown in the following figure:

40

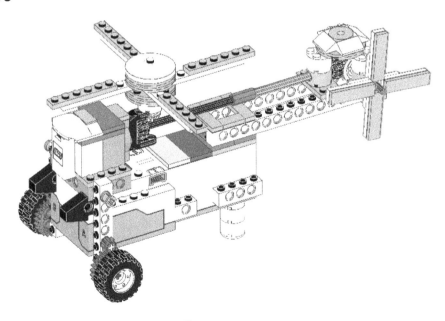

Figure 7.42

41. Take a connector peg and a double-ended tube. Connect the connector peg to the front of the 4x6 brick, and then connect the double-ended tube to it.

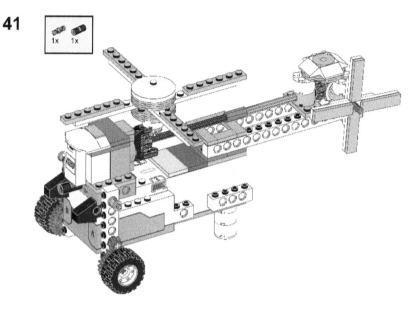

Figure 7.43

42. Take a connector peg with a knob and connect it to the double-ended tube.

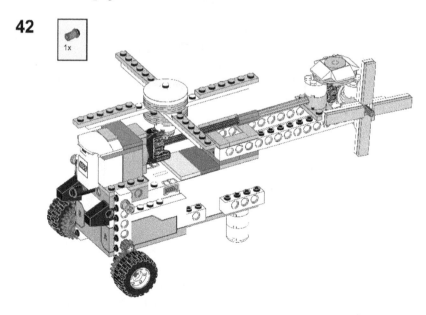

Figure 7.44

43. Take a dome and a container and click them together, as shown here.

 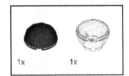

Figure 7.45

44. Now, connect the container and dome to the recently attached connector peg with knob.

44

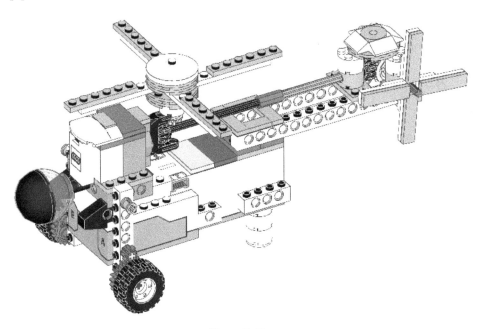

Figure 7.46

The helicopter looks cool, doesn't it? Now, let's solve some fun coding challenges to make this helicopter move and accomplish missions.

Let's code the robot to perform various tasks

Let's start coding this helicopter. Using the extra LEGO bricks from your BOOST kit, build the following:

- A **food element** with orange LEGO bricks

- A **water element** with blue LEGO bricks

- A **rescue equipment element** with yellow LEGO bricks

Make sure that you don't use more than three bricks to build any element. This ensures that the element can be placed on the helicopter with ease and without adding too much load to it.

Activity #1

Program your helicopter to move forward for two rotations. While moving, it should also rotate its wings at 60% power for 3 seconds. You can also add the sound of a helicopter to make it more interesting. The sample code for the activity is given here:

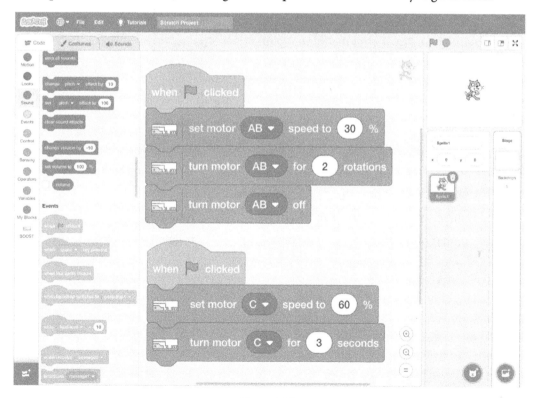

Figure 7.47

As we mentioned previously, a helicopter is used for rescue purposes, military operations, intracity travel, and more. In this activity, we will take our helicopter on a rescue mission. Your helicopter must reach out to three different places where people are stuck with different necessities – food, water, and rescue equipment. You must load these three elements from the base point, go to each of these places, and deliver the necessary items. Exciting, isn't it? Let's do the setup first, as shown in the following diagram. Also, load your helicopter with all three elements before letting it take off.

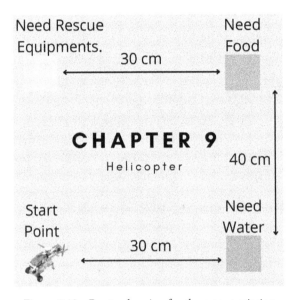

Figure 7.48 – Route planning for the rescue mission

You will have to do some trial-and-error testing to find out the number of rotations for the following:

1. Turning the robot right = _____ rotations.

2. The number of rotations to cover 35 cm = _____ rotations.

3. The number of rotations to cover 40 cm = _____ rotations.

Activity #2

Program your helicopter to go from its starting position and supply water. When you reach the water supply destination, keep your helicopter's wing movement on but stop the movement of the helicopter itself. Wait for 20 seconds so that you can deliver the water safely.

Activity #3

Program your helicopter to reach out to people who need food. Do what you did in the previous activity – make sure that the helicopter stops moving when it reaches the destination but ensure that its wing movement is still on. Wait for 20 seconds so that you can deliver the food safely.

Activity #4

Program your helicopter to reach out to people who need rescue equipment. Do what you did in the previous two activities – make sure that the helicopter stops moving when it reaches the destination but ensure that its wing movement is still on. Wait for 20 seconds so that you can deliver the rescue equipment safely.

Did you know why I asked you to keep the wing movement on? Remember that a helicopter delivers everything while being airborne during such rescue operations. So, the wings should always be moving so that the helicopter can keep its position.

Activity #5

Now, program your helicopter to go back to its base and land.

Time for a challenge

Challenge #1

Take your helicopter on a military operation. You will have to pick up arms and ammunition from two different places and then deliver them to the military base, at an altitude of 3,500 feet.

Ensure that you have the following setup:

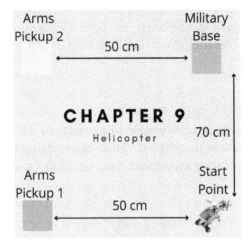

Figure 7.49 – Military operation setup

This challenge shows how helicopters can be useful for rescue and military operations. Similarly, they can be used to reach out to remote places where airplanes cannot land.

Summary

In this chapter, you learned about the application of helicopters and built a helicopter with rotor blades. You coded this helicopter to follow various routes and perform various tasks. You also learned about some important terms such as thrust, drag, lift, and weight.

In the next chapter, we will build the R2D2 robot from Star Wars and go on some exciting missions.

Further reading

You can learn more about helicopters at the following link:

`https://www.nasa.gov/audience/forstudents/5-8/features/nasa-knows/what-is-a-helicopter-58.html`.

You can even research various topics to gain an understanding of helicopters, including the following:

- The most powerful military helicopters in the world
- The top 10 cities in the world where people use helicopters for transport within the city

8
Building R2-D2

Who isn't a Star Wars fan here? As a child, I would make sure to watch the latest offerings from Star Wars on the very first day of release. One of the reasons to love Star Wars is the number of crazy and fancy robots in the movie and the kinds of special actions they do. Of all these robots, I really loved R2-D2 from day 1. Its movement and unique design with a rotating head have always kept me glued to the character. Which is your favorite Star Wars robot? In this chapter, you will be building an R2-D2 robot and solving fun programming tasks:

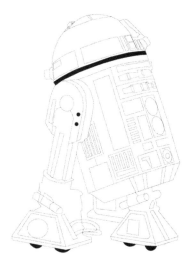

Figure. 8.1 – R2-D2 robot from Star Wars

Let's first understand what a pulley is and how it works. A pulley is one of six simple machines used to make our pulling efforts easy. Some common applications of pulleys are as follows:

- **Curtains/blinds**: A pulley is used to open and close the blinds and helps in pulling the curtains up and down.

- **Flower basket**: A pulley mechanism holds the flower basket, and we can easily lower it when we want to flower the plants.

- **Ropeways**.

- **Rock climbing equipment**.

Like a gear, a pulley system also has a drive and a driven pulley. In your robot, the pulley attached to the neck of R2-D2 will be a driven pulley:

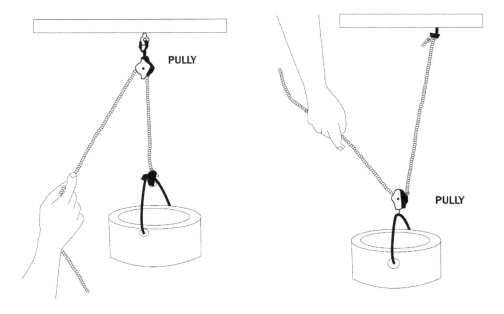

Figure 8.2 – Pulley mechanism (reference image for designer)

This chapter will cover the following areas:

- Building the R2-D2 robot.

- Let's code the robot to move on a specific path

- Time for a challenge.

In this robot, you will be using a single motor drive that will not allow you to make turns. You will be using a pulley mechanism to rotate the head of your R2-D2.

Technical requirements

In this chapter, you will need the following:

- A LEGO BOOST kit with six AAA batteries, fully charged
- A laptop/desktop with Scratch 3.0 programming and an active internet connection
- A diary/notebook with a pencil and eraser

Building the R2-D2 robot

Let's build the following robot:

Figure 8.3 – R2-D2 robot using the BOOST kit

We will use the following building instructions:

1. Take your BOOST Hub. Ensure that the batteries are fully charged:

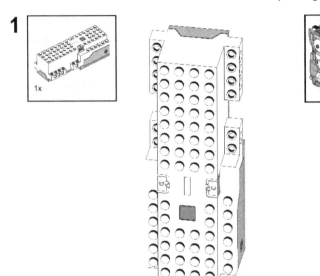

Figure 8.4

2. Take two 1x2 bricks with a horizontal snap and connect them to the BOOST Hub:

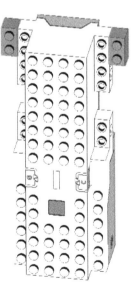

Figure 8.5

3. Take two 2x2 round plates and connect them to the back side of the BOOST Hub as shown in the following figure:

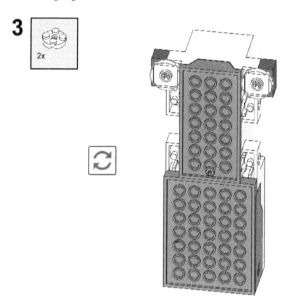

Figure 8.6

4. Now take two 1x6 bricks with a bow and two 1x11 Technic bricks. Connect them to the BOOST Hub, one above the other:

Figure 8.7

5. Take two 2x6 bricks with a bow, one 2M cross axle, and one 1x1 beam. Connect the 2x6 bricks with a bow to the 1x2 bricks and 2M cross axle in motor A of the BOOST Hub, then insert a 1x1 beam:

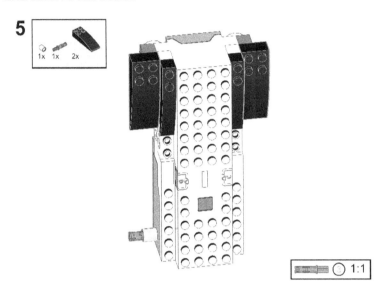

Figure 8.8

6. Take two 1x16 Technic bricks and connect them to the back side of the 2x6 brick with a bow, as shown in the following figure:

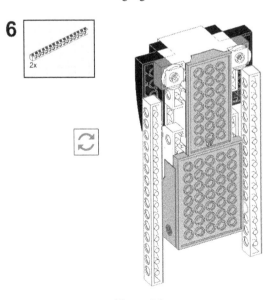

Figure 8.9

7. Take one 3M stopper axle and one 24-tooth spur gear. Connect the spur gear to motor B of the BOOST Hub, then insert the 3M stopper axle onto it:

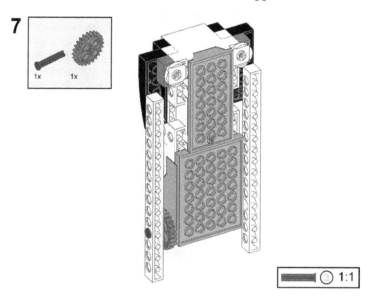

Figure 8.10

8. Take one full bushing, one 5M stopper axle, and one 24-tooth spur gear. Connect the 5M stopper axle to the 15th hole of the Technic brick, then connect the 24-tooth spur gear and full bushing to it, as shown here:

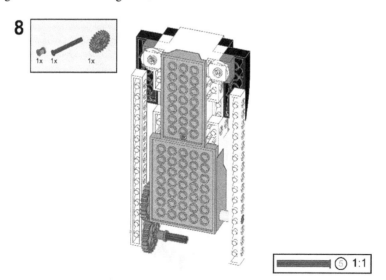

Figure 8.11

9. Take one wheel and connect it to the recently attached 5M stopper axle:

Figure 8.12

10. Take one cross axle extension. Connect it to the recently attached 5M stopper axle after the wheel:

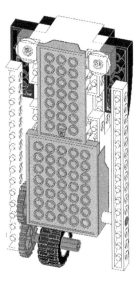

Figure 8.13

11. Take one 3M axle. Connect it to the cross-axle extension:

Figure 8.14

12. Take one wheel and connect it to the recently attached 3M axle:

Figure 8.15

13. Take one connector peg with a cross axle and one cross axle extension. Connect it to the 3M axle as shown in the following figure:

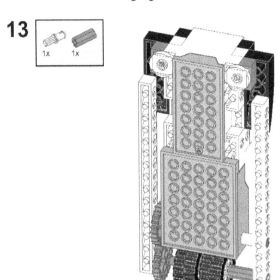

Figure 8.16

14. Take two 1x10 Technic bricks and connect them to the 1x16 Technic brick as shown in the following figure:

Figure 8.17

15. Take two 1x1 beams and two 1x2 roof tiles. Connect them to the 1x16 Technic brick as shown in the following figure:

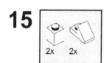

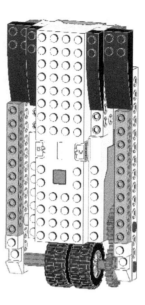

Figure 8.18

16. Now, take two 1x1 plates and two 1x11 plates. Connect the 1x11 plate to the recently attached 1x10 Technic brick and connect the 1x1 plate to the 1x2 roof tile:

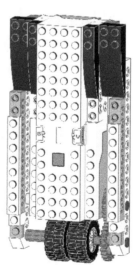

Figure 8.19

17. Take two 1x2x2/3 plates with a bow. Connect them to the recently attached 1x1 plate and 1x1 beam:

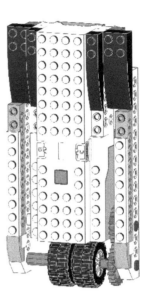

Figure 8.20

18. Now, take two 2x2 round bricks. Connect them to the 2x6 brick with a bow:

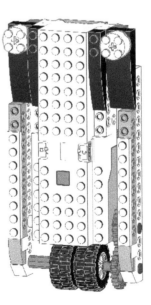

Figure 8.21

19. Take two 1x2 angle plates. Connect them to the top of the BOOST Hub as shown in the following figure:

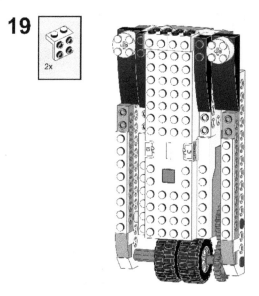

Figure 8.22

20. Now, take one 1x4 plate. Connect it to the back side of the recently attached 1x2 angle plate:

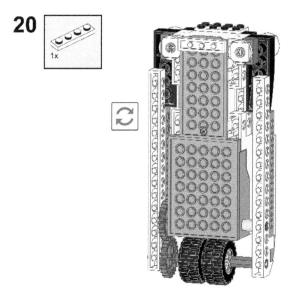

Figure 8.23

21. Take two 1x2 1.5-bot angular plates. Connect them to the recently attached 1x4 plate:

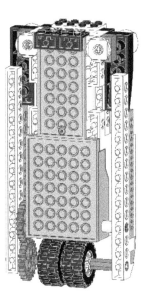

Figure 8.24

22. Take two 2x6 plates. Connect them to the top of the 1x2 1.5-bot angular plate and the 1X2 angle plate as shown in the following figure:

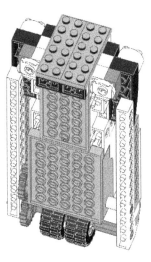

Figure 8.25

23. Take four 1x2 plates. Connect them to the 2x6 plate as shown in the following figure:

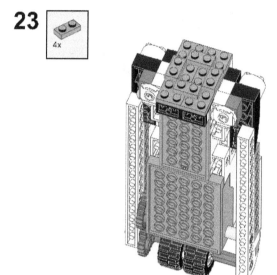

Figure 8.26

24. Take one 3M axle and one 4x4 round plate. Connect the 3M axle to the 4x4 round plate, then connect it to the recently attached 1x2 plate:

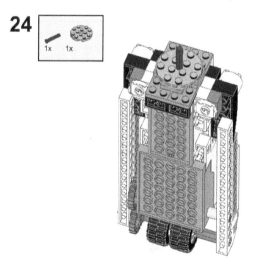

Figure 8.27

25. Take two half bushings and one 2x2 flat tile. First, connect the flat tile to the 2x2 angular plate, and then connect the half bushings through the axle:

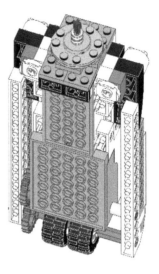

Figure 8 28

26. Take one 3x11x2 left shell and one 3x11x2 right shell. Connect them to the 1x6 brick with a bow as shown in the following figure:

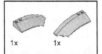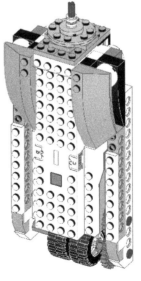

Figure 8.29

27. Take two 1x2 flat tiles and one 1x4 flat tile. Connect the 1x4 flat tile to the 1x2 angle plate and the 1x2 flat tiles to the 3x11x2 left shell and the 3x11x2 right shell:

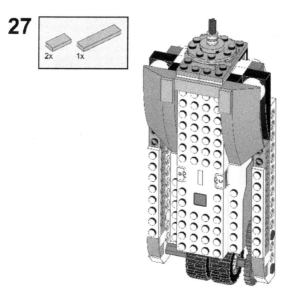

Figure 8.30

28. Now, take one 4x4 round plate. Connect it to the BOOST Hub as shown in the following figure:

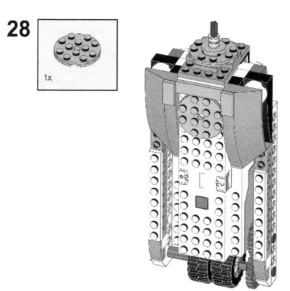

Figure 8.31

29. Take one 3x4x2 plate with a bow and one round plate. Connect the round plate to the 4x4 round plate and the 3x4x2 plate with a bow to the BOOST Hub as shown in the following figure:

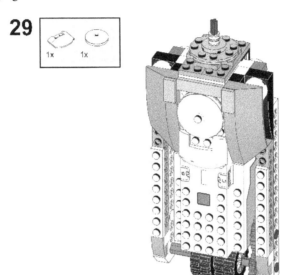

Figure 8.32

30. Take three 1x2 flat tiles. Connect two flat tiles to the 1x11 plate and one flat tile to the 3x4x2 plate with a bow:

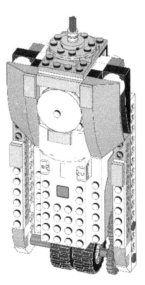

Figure 8.33

31. Now, take six 2x3 plates. Connect them to the BOOST Hub as shown in the following figure:

 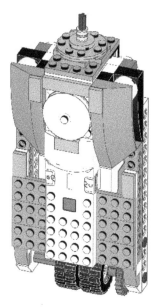

Figure 8.34

32. Now, take three 2x4 bricks with a bow and one 2x4 brick. Connect them to the 2x3 plate as shown in the following figure:

 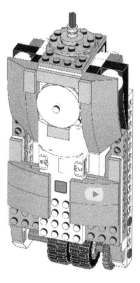

Figure 8.35

33. Take two 1x2 plates. Connect them to the BOOST Hub as shown here:

Figure 8.36

34. Take four 1x2x2/3 plates with a bow. Connect them to the recently attached 1x2 plate as shown in the following figure:

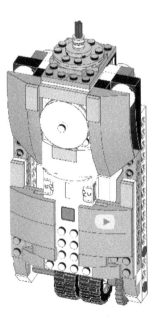

Figure 8.37

35. Take two 1x2 roof tiles. Connect them to the BOOST Hub as shown in the following figure:

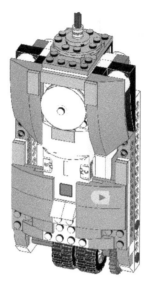

Figure 8.38

36. Take one 2x2x2/3 plate with a bow. Connect it to the recently attached 1x2 roof tile and the BOOST Hub:

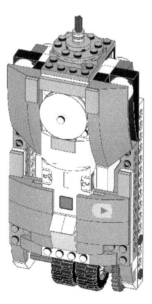

Figure 8.39

37. Take one 1x2 plate. Connect it to the bottom side of the BOOST Hub:

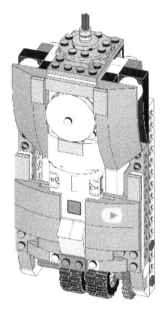

Figure 8.40

38. Take two 1x2x2/3 plates with a bow and connect them to the 1x2 plate:

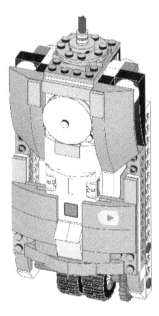

Figure 8.41

39. Now, take two 1x1 orange flat tiles and one 1x1 black flat tile. Connect the orange flat tile to the 2x3 plate and the black flat tile to the round plate:

Figure 8.42

40. Take four connector pegs. Connect them to the first hole and the fifth hole of the 1x16 Technic brick:

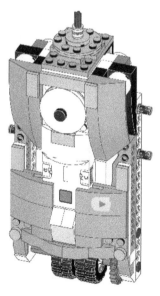

Figure 8.43

41. Now, take one external motor from your BOOST kit. Flip your robot and connect this motor below the 1x2 bricks, as shown in the following figure:

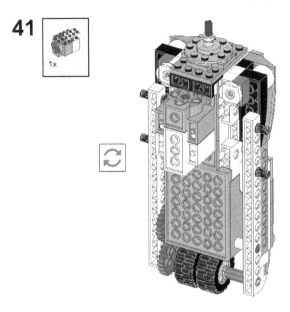

Figure 8.44

42. Take two 3M connector pegs. Connect them to the external motor:

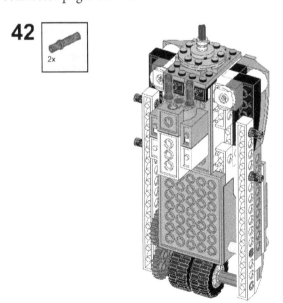

Figure 8.45

43. Take two 3M Technic beams. Connect them to the recently attached 3M connector peg:

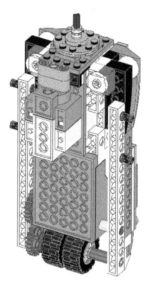

Figure 8.46

44. Take two half bushings and one 4M stopper axle. Insert the half bushing on the 4M stopper axle, then connect the 4M stopper axle to the external motor through the 3M beam:

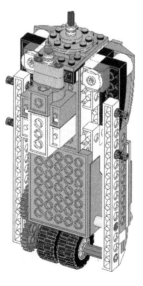

Figure 8.47

45. Take one rubber band. Connect it to the half bushing as shown in the following figure:

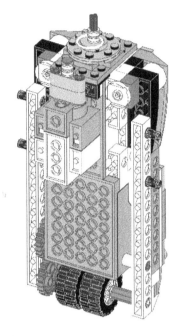

Figure 8.48

46. Now, stretch the rubber band and connect it to the half bushing of the external motor:

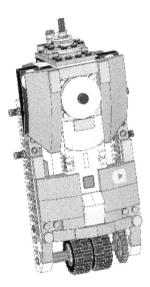

Figure 8.49

47. Now, take one 2x2 round brick:

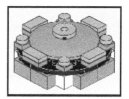

Figure 8.50

48. Take four 4x4 ¼-circle plates. Arrange them as shown in the following figure:

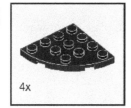

Figure 8.51

49. Take one 4x6 brick. Connect it to the back side of the ¼-circle plate and connect the 2x2 round brick to the center of the ¼-circle plate:

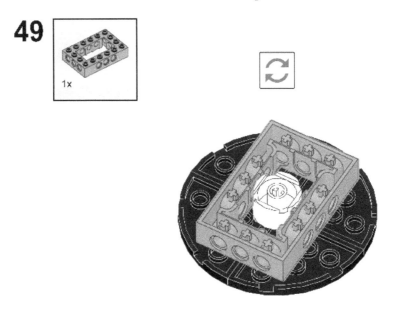

Figure 8.52

50. Take four 1x1 bricks. Connect them to the ¼-circle plate as shown in the following figure:

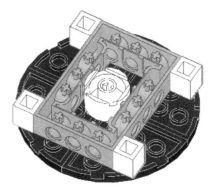

Figure 8.53

51. Take four 2x2 roof tiles. Connect them to both sides of the ¼-circle plate:

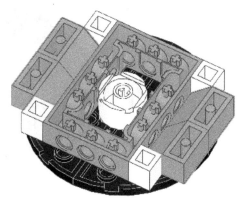

Figure 8.54

52. Take four 1x2 bricks and connect them to the ¼-circle plate as shown in the following figure:

Figure 8.55

53. Take one 4x4 round plate. Connect it to the top side of the ¼-circle plate:

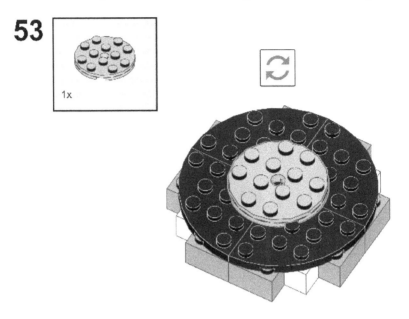

Figure 8.56

54. Take one 2x2 round plate. Connect it to the 4x4 round plate:

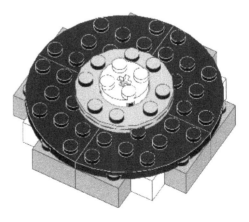

Figure 8.57

55. Take one 6x6 parabola. Connect it to the 2x2 round plate as shown in the following figure:

Figure 8.58

56. Take one 2x2 flat round tile. Connect it to the 6x6 parabola:

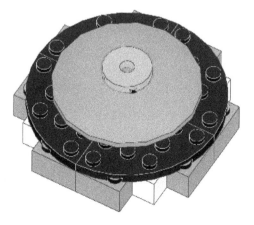

Figure 8.59

57. Take four 1x2 plates. Connect them to all sides of the ¼ round plate as shown in the following figure:

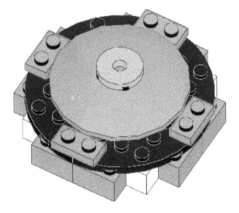

Figure 8.60

58. Take four 1x1 plates. Connect them to the middle of the ¼ round plate as shown in the following figure:

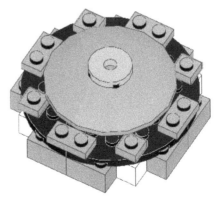

Figure 8.61

59. Take four 1x2 flat tiles. Connect them to the 1x2 plate as shown here:

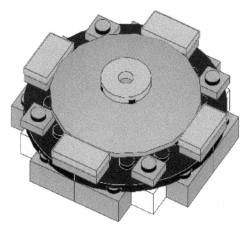

Figure 8.62

60. Take four 1x1 flat round tiles. Connect them to the 1x1 plate:

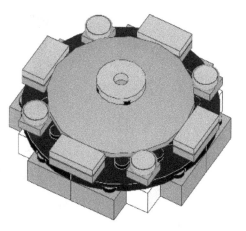

Figure 8.63

61. Take the R2-D2 head and align it with the 3M axle:

61

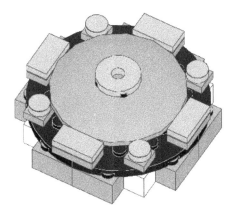

Figure 8.64

62. Now, connect it to the 3M axle as shown in the following figure:

62

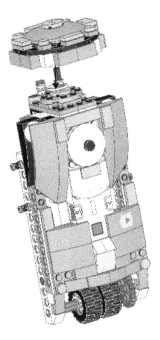

Figure 8.65

63. After connecting that part, it will look as shown in the following figure:

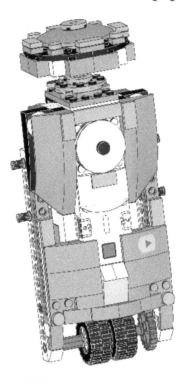

Figure 8.66

64. Take one 1x16 Technic brick:

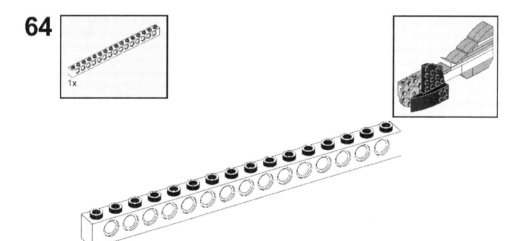

Figure 8.67

65. Take one 1x6 brick. Hold the 1x6 brick beside the 1x16 Technic brick:

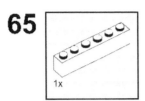

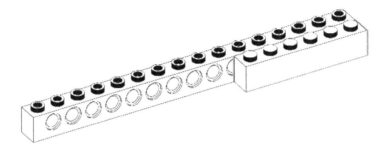

Figure 8.68

66. Take two 2x2 roof tiles and one 2x2 brick with a horizontal snap. Flip the recently connected Technic bricks and connect these new pieces to them as shown in the following figure:

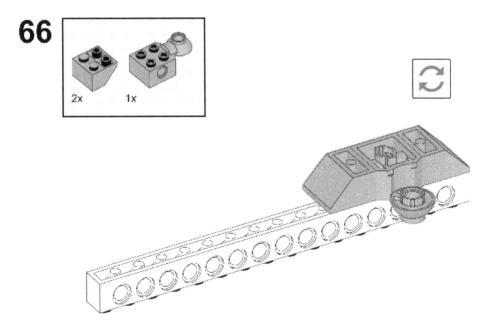

Figure 8.69

67. Take one 2x11 plate with holes. Connect it to the top of the 1x16 Technic brick:

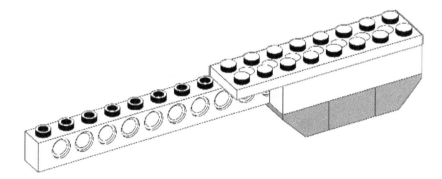

Figure 8.70

68. Take one 1x4 brick and connect it below the 2x11 plate as shown in the following figure:

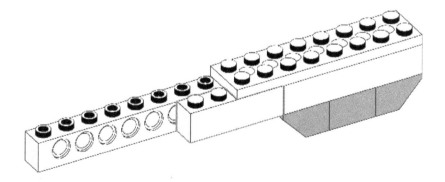

Figure 8.71

69. Take one 2x2 plate and one 2x4 plate. Connect them to the 2x11 plate with a hole:

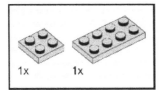

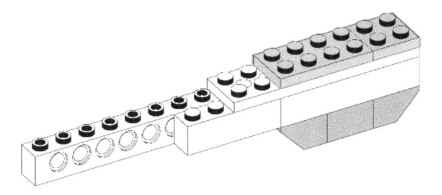

Figure 8.72

70. Take one 1x2 plate and one 2x4 plate. Connect the 1x2 plate to the 2x2 plate and connect the 2x4 plate to the 2x4 plate and the 2x2 plate:

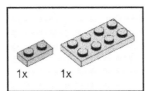

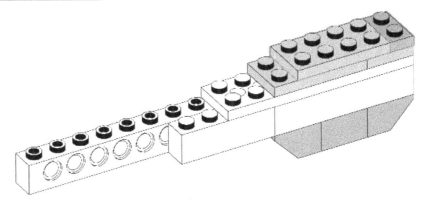

Figure 8.73

71. Take one 2x4 plate and connect it to the stack of plates as shown here:

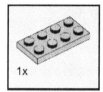

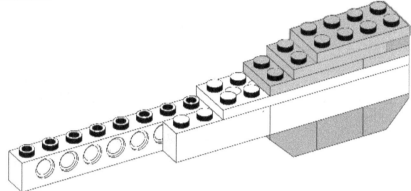

Figure 8.74

72. Take one 2x3 plate and connect it to the 2x4 orange plate, then take one 1x2 plate and connect it over them:

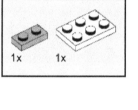

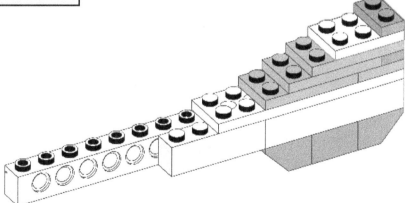

Figure 8.75

73. Take two 1/3 bricks with a bow and connect them to the stack of plates as shown in the following figure:

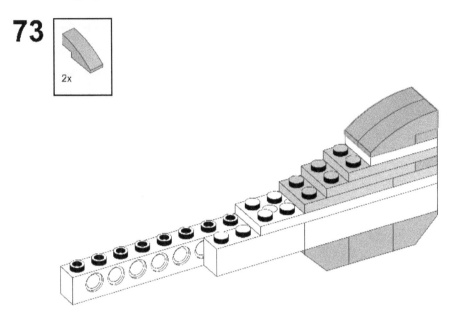

Figure 8.76

74. Take two 1x2x2/3 plates with a bow and connect them as shown in the following figure:

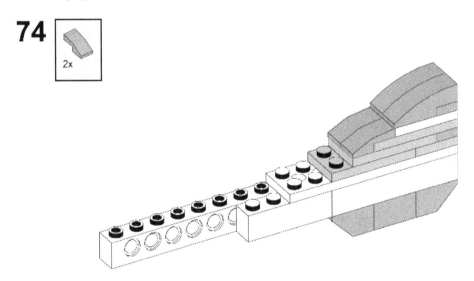

Figure 8.77

75. Once again, take two 1x2x2/3 plates with a bow and connect them as shown here:

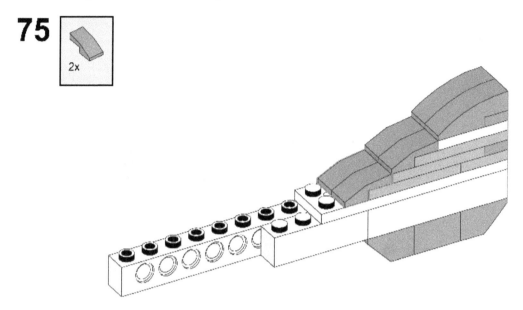

Figure 8.78

76. Take one 1x2x2/3 roof tile and connect it as shown in the following figure:

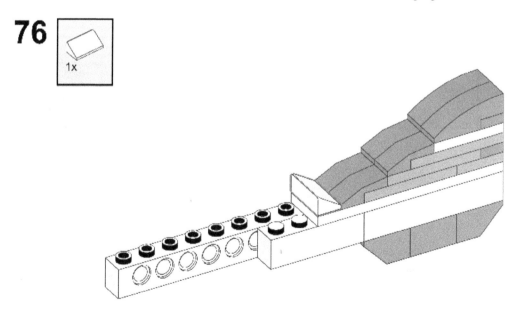

Figure 8.79

77. Take one 2x2 plate with a knob and connect it to the white bricks:

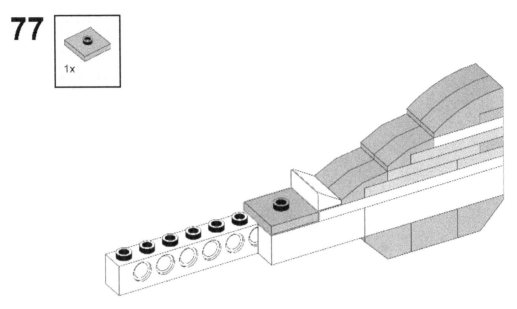

Figure 8.80

78. Now, take one 1x4 plate and one 1x2/2x2 angle plate and connect them as shown in the following figure:

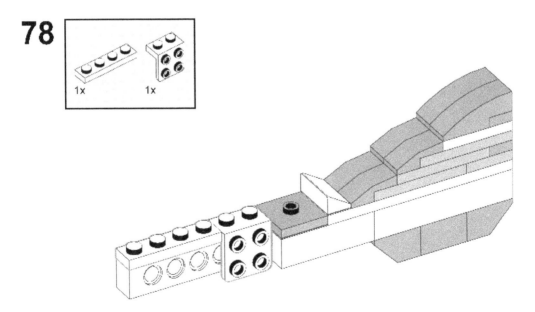

Figure 8.81

79. Take one 1x2 plate and one 1x4 plate with two knobs and connect them to the arm as shown in the following figure:

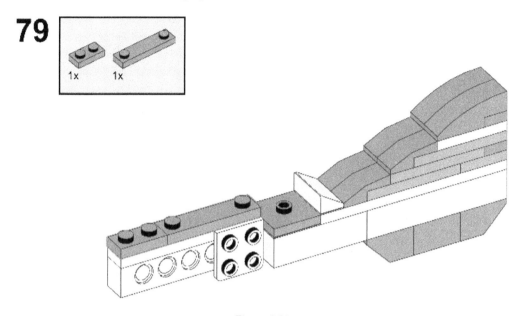

Figure 8.82

80. Take one connector peg and connect it to the brick as shown in the following figure:

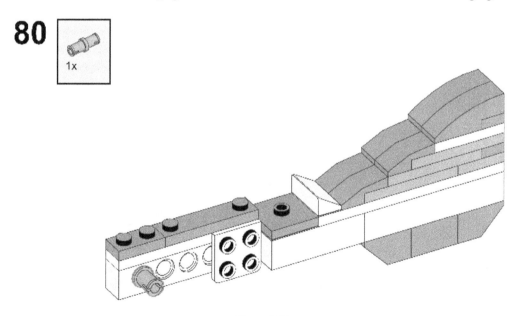

Figure 8.83

81. Take one 2x2 flat round tile with a hole and one 4x4 round plate with a snap and connect them as shown in the following figure:

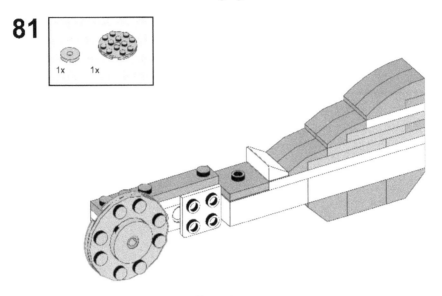

Figure 8.84

82. Take one 2x4 brick and connect it to the 1x2/2x2 angle plate:

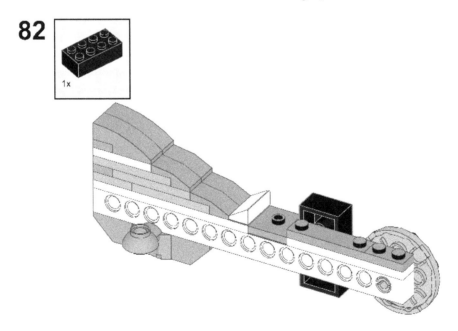

Figure 8.85

83. Take one 2x6 brick with a bow and one ¼ circle of the 4x4 plate and connect them to the 2x4 brick:

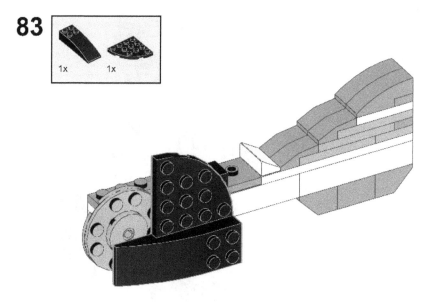

Figure 8.86

84. Now connect this arm to the body of R2-D2, as shown in the following figure:

Figure 8.87

85. Check that the connection is correct by comparing it with the following figure:

85

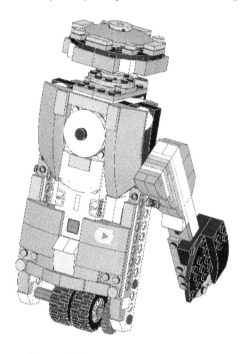

Figure 8.88

86. Now, we are going to make the other arm for our R2-D2. Let's start by taking one 1x16 Technic beam:

86

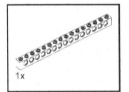

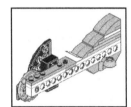

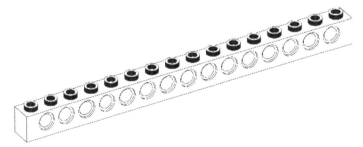

Figure 8.89

87. Take one 1x6 brick and put it beside the 1x16 brick:

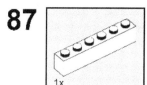

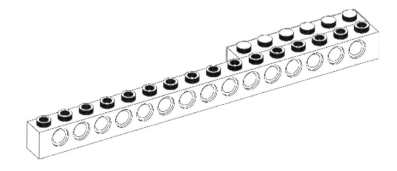

Figure 8.90

88. Take one 2x2 brick with a horizontal snap and connect it to the back side of both bricks as shown. Then, take two 2x2/45 inverted roof tiles and connect them to both sides of the 2x2 brick with a horizontal snap:

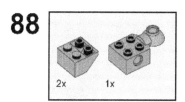

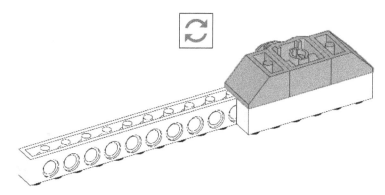

Figure 8.91

89. Take one 1x4 brick and put it in place as shown:

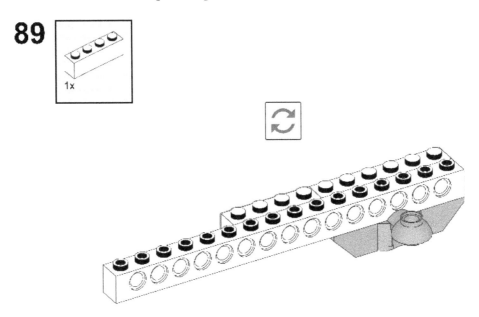

Figure 8.92

90. Take one 2x11 plate. Connect the 1x16 brick to the 1x6 and 1x4 bricks using this plate as shown in the following figure:

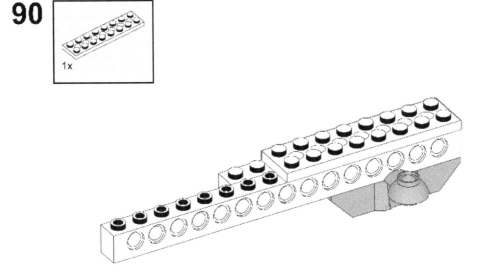

Figure 8.93

91. Take one 2x2 plate and one 2x4 plate and attach them to the 1x11 plate:

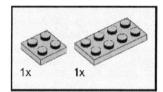

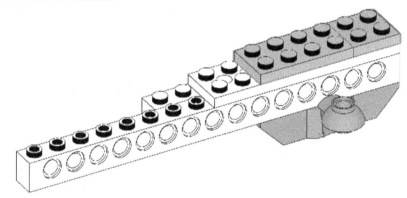

Figure 8.94

92. Now, take one 1x2 plate and connect it to the 2x2 plate and take one 2x4 plate and connect it to both the 2x2 and 2x4 plates, as shown in the following figure:

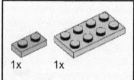

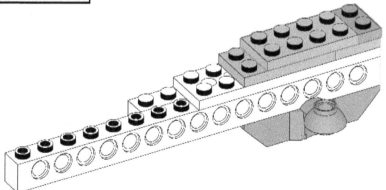

Figure 8.95

93. Take one 2x4 plate and connect it to the stack of plates:

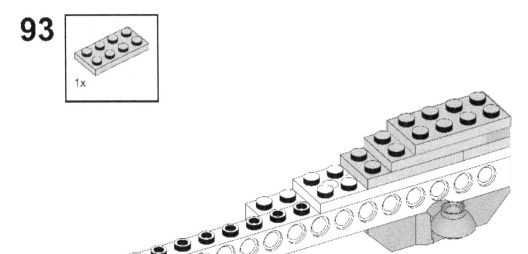

Figure 8.96

94. Take one 2x3 plate and connect it to the stack of plates:

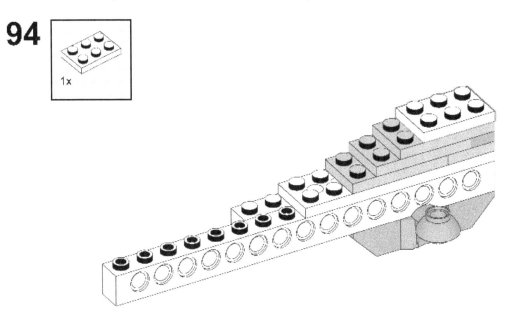

Figure 8.97

95. Take one 1x2 plate and connect it as shown:

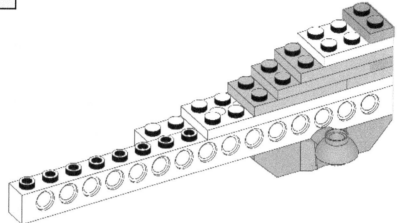

Figure 8.98

96. Take two 1/3 bricks with a bow and connect them to the stack of plates as shown in the following figure:

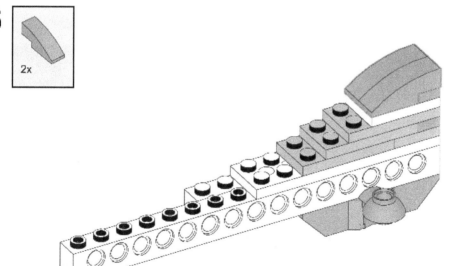

Figure 8.99

97. Take two 1x2x2/3 plates with a bow and connect them as shown in the following figure:

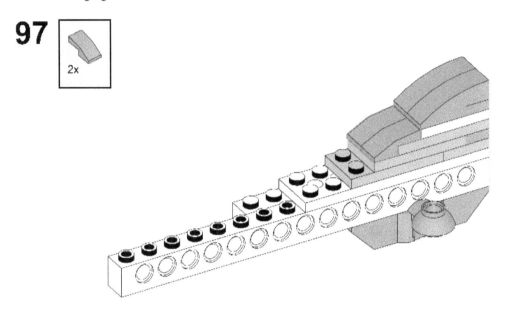

Figure 8.100

98. Once again, take two 1x2x2/3 plates with a bow and connect them as shown here:

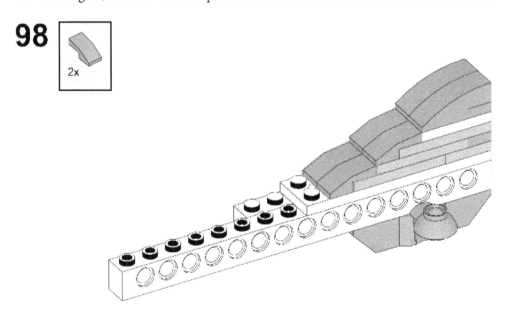

Figure 8.101

99. Take one 1x2x2/3 roof tile and connect it as shown in the following figure:

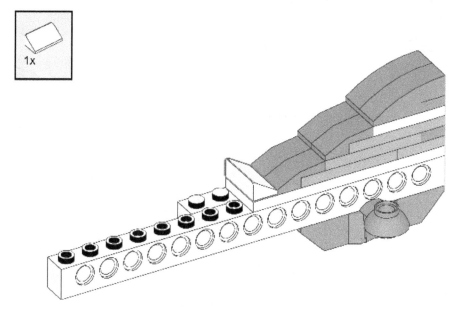

Figure 8.102

100. Take one 2x2 plate with a knob and connect it to the white bricks:

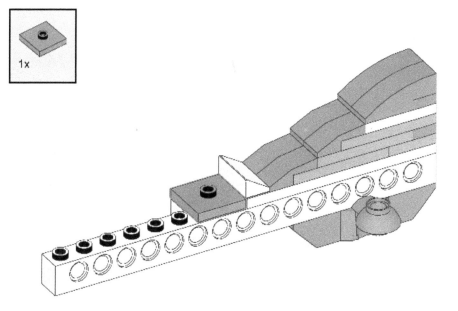

Figure 8.103

101. Now take one 1x4 plate and one 1x2/2x2 angle plate and connect them as shown in the following figure:

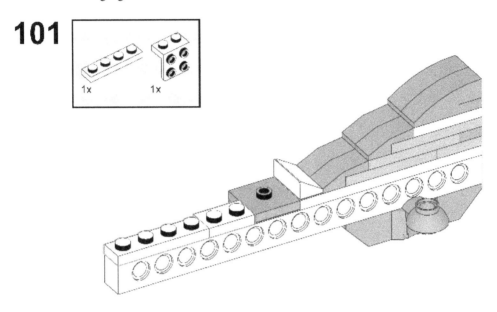

Figure 8.104

102. Take one 1x4 plate with two knobs and connect it to the arm as shown in the following figure:

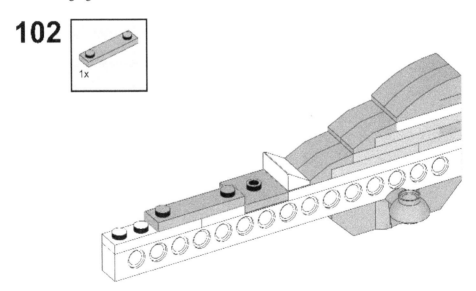

Figure 8.105

103. Take one connector peg and connect it to the brick as shown in the following figure:

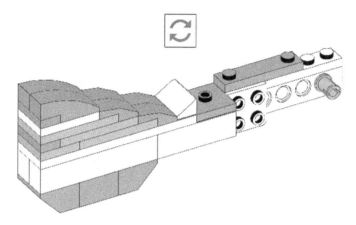

Figure 8.106

104. Take one 2x2 flat round tile with a hole and one 4x4 round plate with a snap and connect them as shown in the following figure:

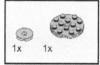

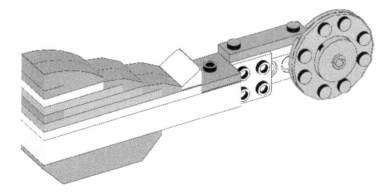

Figure 8.107

105. Take one 1x2 plate and connect it to the arm as shown in the following figure:

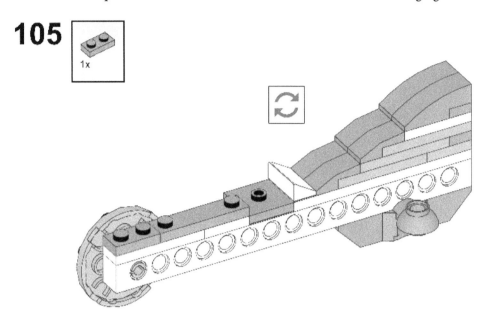

Figure 8.108

106. Take one 2X4 brick and connect it to the 1x2/2x2 angle plate:

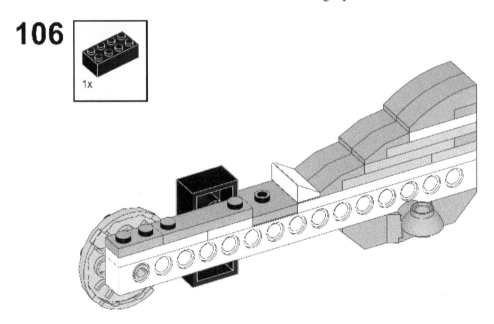

Figure 8.109

107. Take one 2x6 brick with a bow and one ¼ circle of the 4x4 plate and connect them to the 2x4 brick:

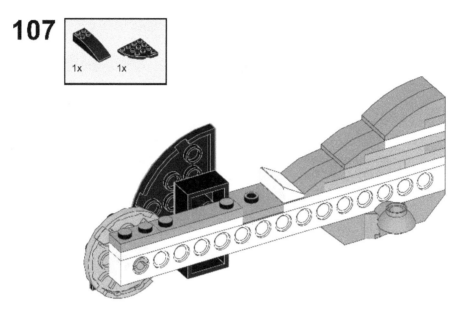

Figure 8.110

108. Compare your model to the model shown in the following figure:

108

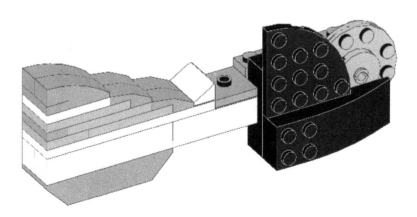

Figure 8.111

109. Now connect this arm to the other side of the body of R2-D2 as shown:

109

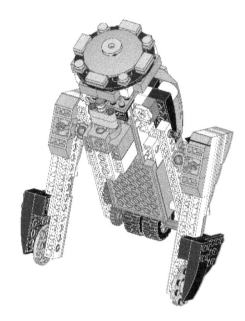

Figure 8.112

110. Check that the connection is correct by comparing it to the following figure:

110

Figure 8.113

111. Take one 2x12 plate and connect it to the back side of the model to support the arms:

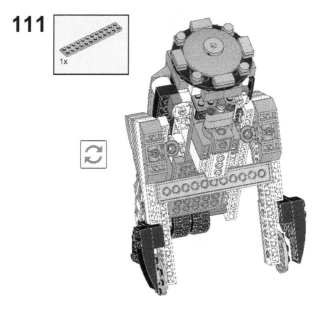

Figure 8.114

112. Once again, take one 2x12 plate and connect it below the first one:

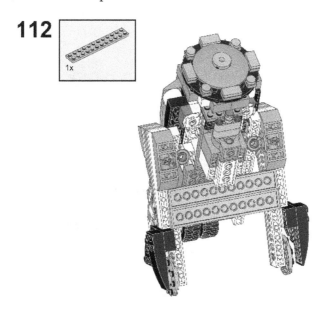

Figure 8.115

And there you go! Your R2-D2 lookalike robot using the BOOST kit is ready:

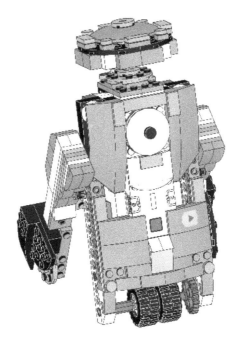

Figure 8.116

Let's perform some fun tasks using this R2-D2 robot. Since the drive wheels of this robot are connected to a single motor, it will not be able to make turns.

Let's code the robot to move on a specific path

In this activity, you will be making use of sound and display blocks to make the R2-D2 more interactive.

Activity

Do the setup as shown in *Figure 8.117*. R2-D2 is on a rescue mission for his master – Padme. R2-D2 needs to reach the computer system in under 5 seconds and turn it off by moving his head 180 degrees after reaching the spot. Once done, display **Padme Saved! Mission Successful** and play *Hooray*:

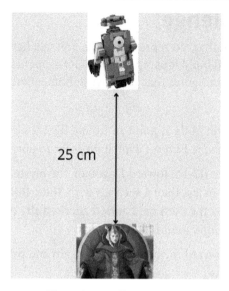

Figure 8.117 – Route setting

The sample code for this activity is given here:

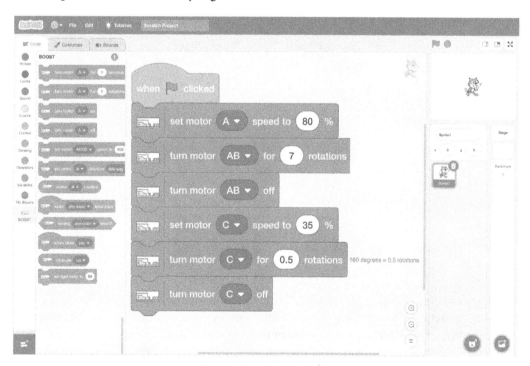

Figure 8.118 - Sample code

Now, let's move on to a challenge.

Time for a challenge

Build a small LEGO element that can represent Yoda. You will have to attach this element to your R2-D2 robot. You can build it using spare LEGO elements from your BOOST kit/use any of your LEGO minifigures that you have at home. Once done, program your R2-D2 to do the following:

1. Go and meet Jedi Master Yoda at point A. Rotate R2-D2's head one time when the task is done and change the BOOST Hub light color to purple.

2. Reorient the position of R2-D2 toward Dagobah, the mysterious force planet. Now, reach this planet in less than 6 seconds with Yoda. Rotate the neck for one full rotation and display the message **Landed successfully on Dagobah with the master**. Also play a unique sound for 3 seconds.

3. Now, reorient R2-D2 toward the base position again and program it to reach the base in under 4 seconds.

4. Calculate the total time taken to perform this activity once you are set with the route. Time taken = _____ seconds.

Your route setting will look as shown in the following figure:

Figure 8.119 – Route for the challenge

You can try creating different routes and solving different missions for R2-D2 for your own coding practice. You will have to orient the robot in the direction you need to solve the mission since this R2-D2 is not capable of making turns. You can even consider making modifications to this construction and drive both wheels with different motors.

Summary

In this chapter, you built an R2-D2 robot using the application of gears as well as a simple machine pulley. You solved two tasks: one to rescue Padme and another to pick up the Jedi Master and go to the planet Dagobah. The purpose of this chapter was to strengthen your construction skills using the pieces given in your LEGO BOOST kit. By now, you should be able to build robots that have human-like movements, such as the neck movement of R2-D2. You should also be able to design robots that are like real-life/sci-fi character robots that you have seen either onscreen or in real life. In the next chapter, you will be building an automated entrance door with the help of a color sensor. The door construction will resemble the entrance that you see at malls/offices where the door will open only for specific people and not for others. It will be a fun task for you.

Further reading

- You can learn more about the features and importance of R2-D2 at `https://www.starwars.com/databank/r2-d2`.

- To learn more about pulleys, visit `https://www.flight-mechanic.com/simple-machines-the-pulley/`.

9
Building an Automatic Entrance Door

Automatic entrance doors are used at various places, such as malls, offices, and so on. At malls, they ensure contactless entry and timely opening and closing of the door, with no human interaction required. At offices, such doors are combined with employees' **identifier** (**ID**) cards and these doors open only when the right ID card is placed in front of the door sensor, or the door stays closed. This ensures the safety of all employees in the office premises as well as helping with attendance management. In this project, you will be applying the concept of a simple machine pulley and building an entrance door that only opens when certain colors are detected. You might have seen automatic doors at malls, offices, airports, metro stations, and other places in your everyday life. Usually, an **infrared** (**IR**) sensor is used to detect the presence of a human and open/close the door. In our case, we will use a color sensor.

Do you know that distance sensors either use ultrasonic sound or IR waves to detect obstacles? They have a transmitter that emits waves, which are then reflected by an obstacle that is received by the receiver. The time difference between transmission and reception of the waves is noted, and the distance is measured using a basic formula.

You can see an automatic entrance door installation in the following screenshot:

Figure 9.1 – Automatic entrance door installation

In this chapter, you will do the following:

- Building an automatic entrance door
- Let's code the door to open under certain conditions
- Time for a challenge

Technical requirements

In this chapter, you will need the following:

- LEGO BOOST kit with six AAA batteries, fully charged
- Laptop/desktop with Scratch 3.0 programming installed and an active internet connection
- A diary/notebook with pencil and eraser

Building an automatic entrance door

Let's build a door by following the building instructions given next. This is what our door should eventually look like:

Figure 9.2

We'll proceed as follows:

1.   Let's start our construction by taking two 1x10 bricks to make the base, as
    illustrated here:

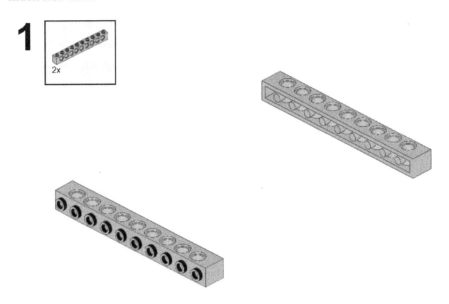

Figure 9.3

2.   Then, take six connector pegs and connect them to the bricks, like so:

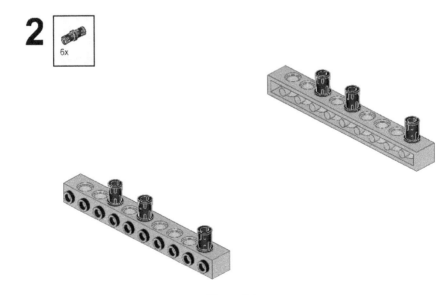

Figure 9.4

3. Now, take two 1x16 bricks and connect them to each other, as follows:

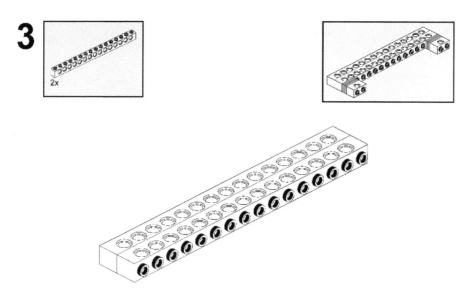

Figure 9.5

4. Then, take four 1x2 plates and connect them to both ends of the brick, as follows:

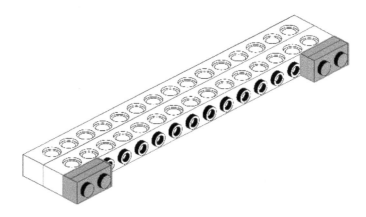

Figure 9.6

5. Take two 1x2 bricks and connect them to the plates, like this:

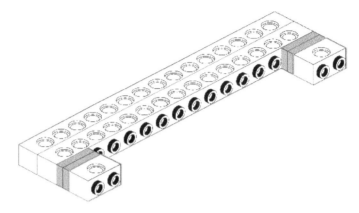

Figure 9.7

6. Now, mount this whole structure onto the blue bricks with connector pegs. Then, take two tubes with double holes and connect them with connector pegs, as illustrated here:

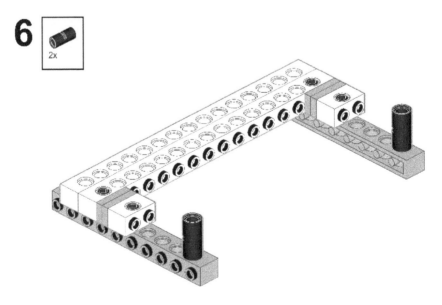

Figure 9.8

7. Take two connector pegs and two 2M cross axles with snap, and connect them to the brick, as shown here:

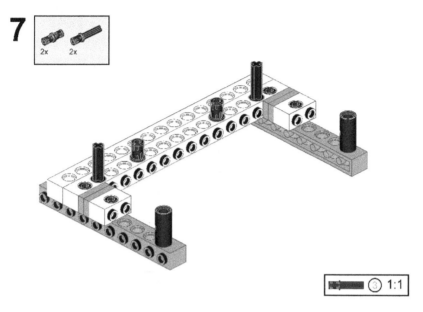

Figure 9.9

8. Take two 1x2 bricks with cross holes and attach them with both the axles. Then, take one 7M beam and connect it with both the connector pegs, as shown here:

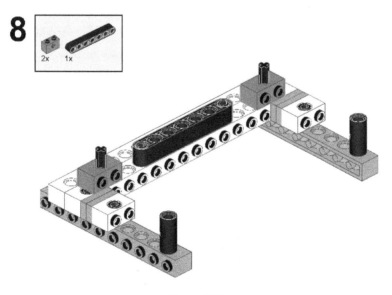

Figure 9.10

9. Take two 2M cross axle connectors, connect them as shown, and then take two 7M cross axles and connect them with both the connector pegs, as follows:

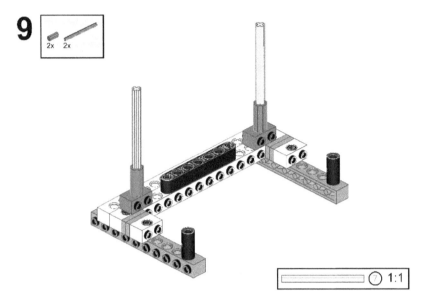

Figure 9.11

10. Take two 180°-angle elements and connect them with both the axles, then take two 4M axles and connect them, as shown here:

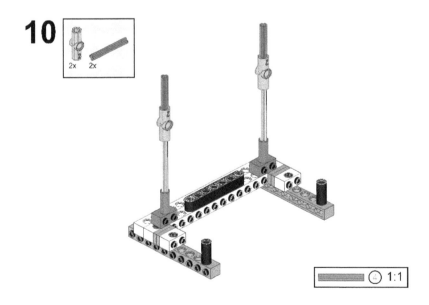

Figure 9.12

11. Take two 1x2 bricks with cross holes and connect them with the axle, as shown here:

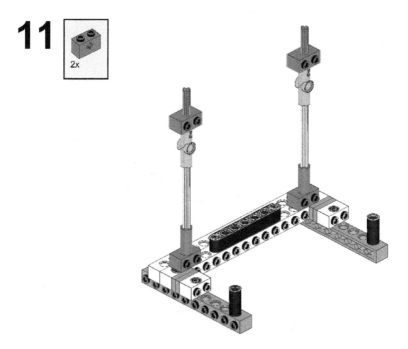

Figure 9.13

12. Now, take one 7M beam and connect two connector pegs to both the ends of the beam, as follows:

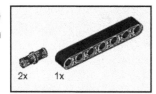
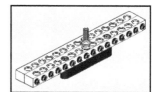

Figure 9.14

13. Take one 1x16 brick and connect it to the beam, as shown here:

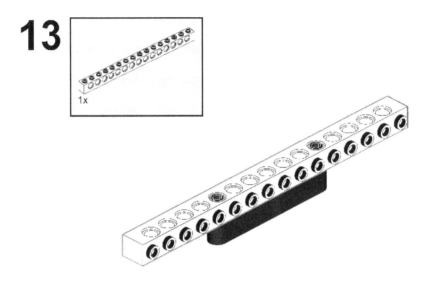

Figure 9.15

14. Take one 4M stop axle and connect it through the 1x16 brick as well as the 7M beam, as shown in the following figure:

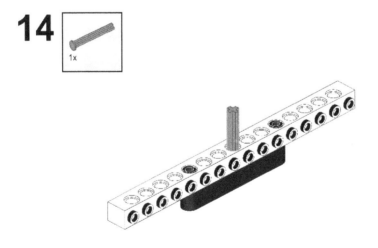

Figure 9.16

15. Take two ½ bushes and attach them with the 4M stop axle, as follows:

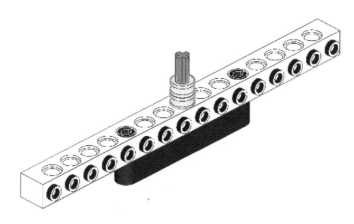

Figure 9.17

16. Now, take one 1x16 brick and connect it, as shown here:

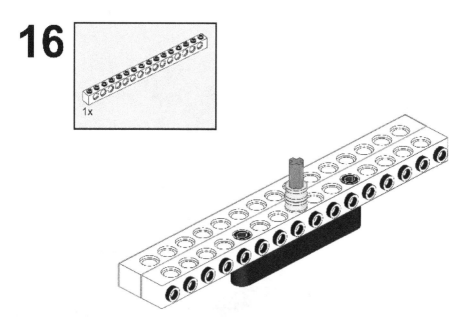

Figure 9.18

17. Now, connect this whole structure with red colored axles, as shown here:

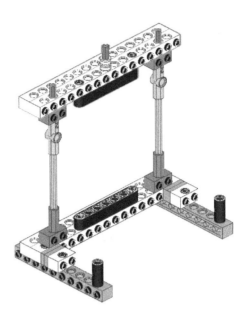

Figure 9.19

18. Take 4x ½ bushes and connect them with the axles, as shown here:

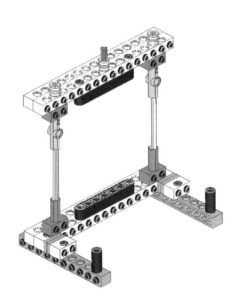

Figure 9.20

19. Take two 2x12 plates and connect them to the 1x2 bricks, as shown here:

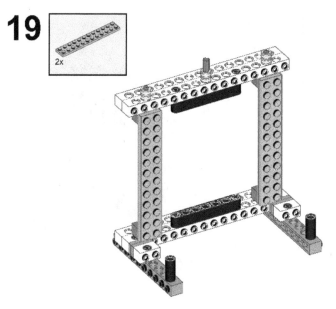

Figure 9.21

20. Now, take four 2x8 plates and connect them horizontally to the 2x12 plates as shown, then take two 2x6 plates and connect them vertically to both the 2x12 plates, as follows:

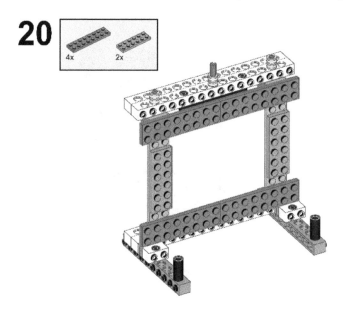

Figure 9.22

21. Take two 6x10 plates and connect them, as shown here:

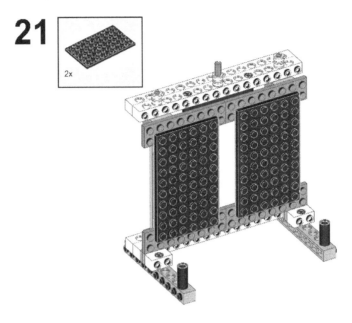

Figure 9.23

22. Take four 1x10 plates and connect them, as shown here:

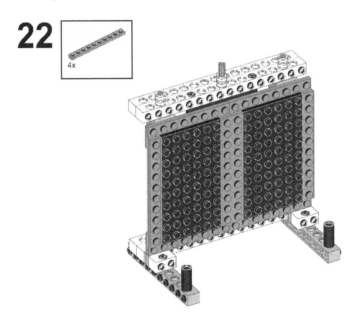

Figure 9.24

23. Take four 1x8 plates and connect them to the top and bottom of the door, like this:

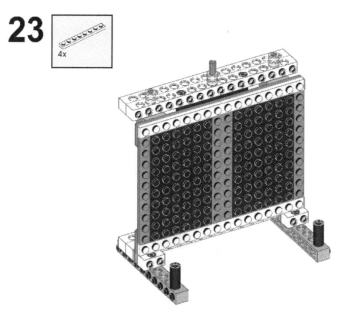

Figure 9.25

24. Take four 4x4 plates and connect them to the down side of the door, as follows:

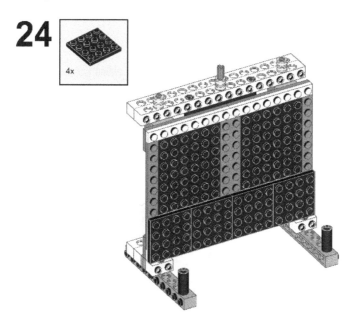

Figure 9.26

25. Now, take two 2x4 plates and connect them to the top of the door, as shown. Then, take two 3x8 left plates with angles and connect them to the top-right and top-left corner of the door, respectively, as illustrated here:

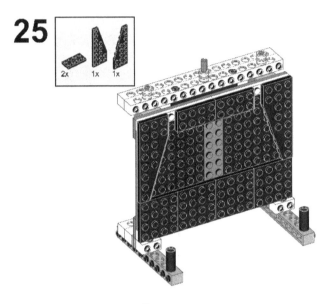

Figure 9.27

26. Then, take two 1x1 plates and connect them to the door, as shown here:

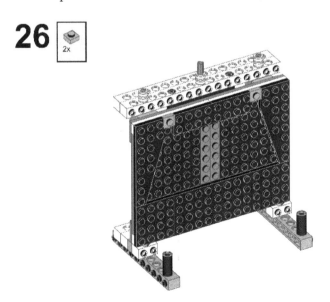

Figure 9.28

27. Take four 1x4 plates and connect them to the door, as shown here:

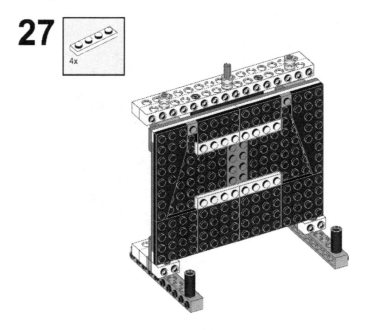

Figure 9.29

28. Then, take two 1x4 plates and connect them to the door, as shown here:

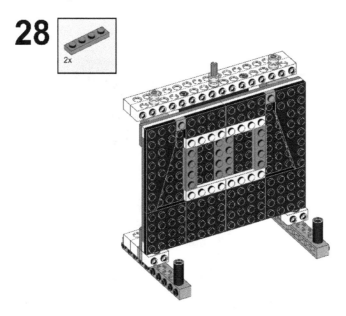

Figure 9.30

29. Take two 1x8 flat tiles and connect them, as shown here:

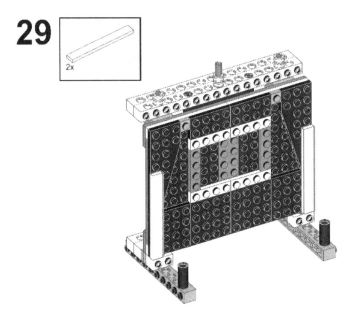

Figure 9.31

30. Then, take four 1x3 flat tiles and connect them to the top-left and top-right corner of the door, as shown here:

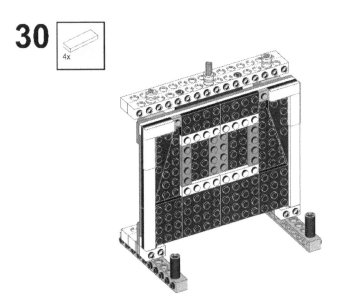

Figure 9.32

31. Now, take four 1x4 flat tiles and connect them to the door, as shown here:

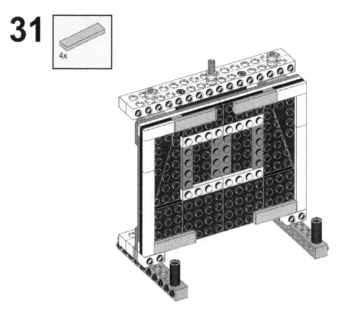

Figure 9.33

32. Now, take four 1x2 flat tiles and connect them, as shown here:

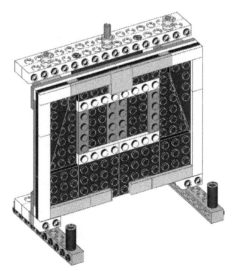

Figure 9.34

33. Take two 1x4 flat tiles and connect them to the bottom edges of the door, as shown here:

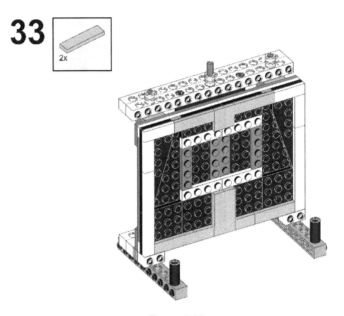

Figure 9.35

34. Take four 1x4 flat tiles and connect them to the door, as shown here:

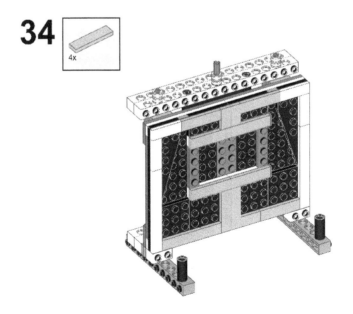

Figure 9.36

35. Take two 1x4 flat tiles and two 1x4 plates with two knobs, and connect them as shown here:

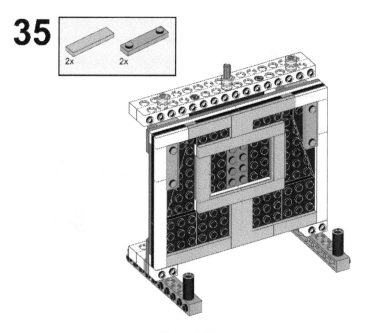

Figure 9.37

36. Now, let's make handles for the door. To do that, take two 2M axles, as illustrated here:

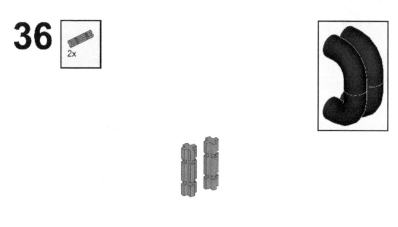

Figure 9.38

37. Then, take two design shapes with tubes and cross holes and connect them to the axle, as shown here:

Figure 9.39

38. Once again, take two design shapes with tubes and cross holes and connect them, as shown here:

Figure 9.40

39. Then, connect both the handles to the door, as shown here:

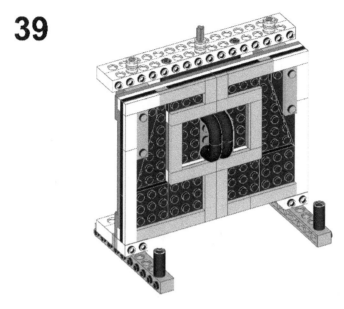

Figure 9.41

40. Take two white rubber bands and connect them with half bushes, as shown here:

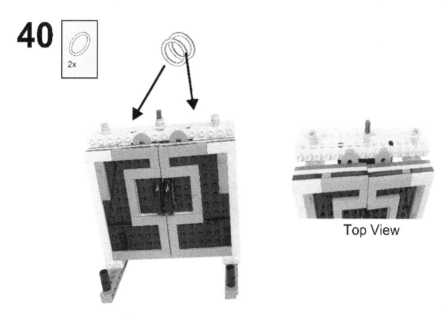

Top View

Figure 9.42

41. Now, take an external motor from your BOOST kit and connect two 1x2 plates to the backside of the motor, as shown here:

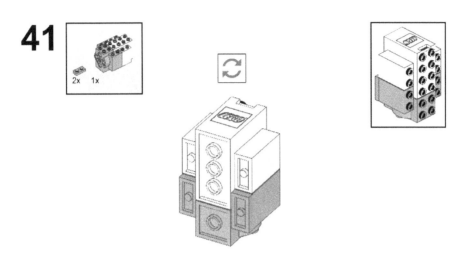

Figure 9.43

42. Then, connect the motor to the central axle of the door, as shown here:

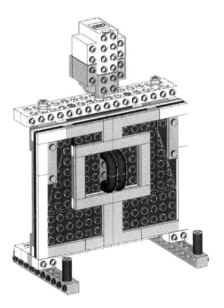

Figure 9.44

43. Now, let's make it automated by attaching a sensor to it. To do that, take a 2x2 plate, as illustrated here:

Figure 9.45

44. Take two 1x2 plates with ball ends and connect them to the 2x2 plate, as follows:

Figure 9.46

45. Take two 1x2 plates with ball cups and connect them to the ball-end plates, as follows:

Figure 9.47

46. Take one 2x2 plate and connect it, as shown here:

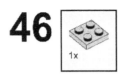

Figure 9.48

47. Now, take a color sensor and connect it to the plates, as shown here:

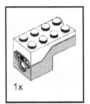

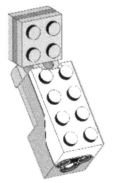

Figure 9.49

48. Connect the whole structure to the external motor, as shown here:

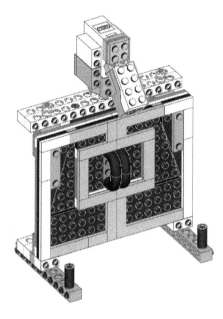

Figure 9.50

49. Now, we're going to connect the BOOST Hub to the model. For that, take the hub from your kit. You can see this illustrated here:

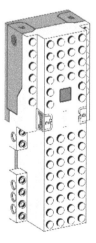

Figure 9.51

50. Take two 1x2 plates and connect them to the BOOST Hub, as shown here:

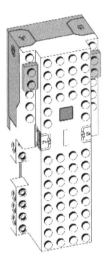

Figure 9.52

51. Then, take two 1x6 plates and connect them to the 1x2 plates, as shown here:

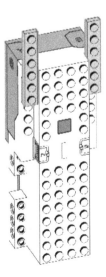

Figure 9.53

52. Take one 2x6 plate and connect it to the orange plates, like this:

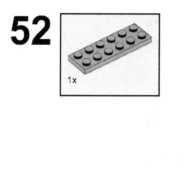

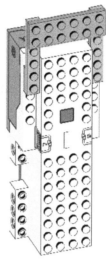

Figure 9.54

53. Now, connect this structure to the motor and the base of the main model, as follows:

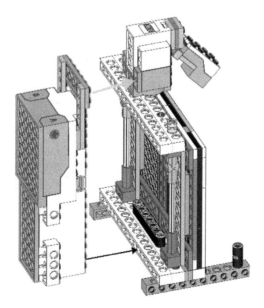

Figure 9.55

54. Check the connections. If correctly connected, your model should look like this:

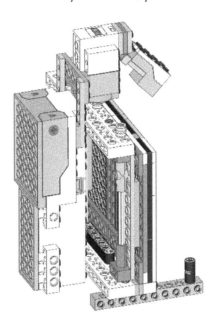

Figure 9.56

Compare your model with the model shown here:

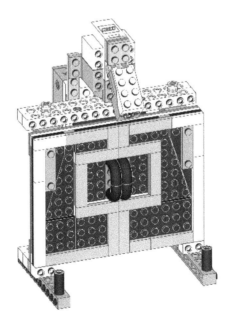

Figure 9.57

Let's now build a man to stand in front of this gate with different colored hats, using the spare LEGO bricks from your BOOST kit, as follows:

1. Take two 2x6 bricks with bows and connect them, as shown here:

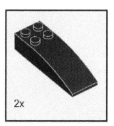

Figure 9.58

2. Now, take two 2x4 plates with holes and connect them to the 2x6 bricks with bows, as follows:

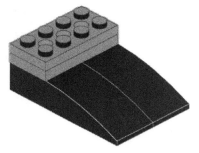

Figure 9.59

3. Take two connector pegs and connect them to the first and third holes of the upper 2x4 plate, as follows:

Figure 9.60

4. Take two connector pegs and one 4x6 brick. Connect the 4x6 brick to the recently attached connector peg, and on the 4x6 brick, connect two connector pegs, as illustrated here:

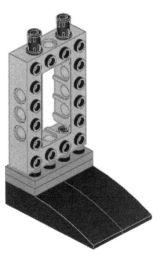

Figure 9.61

5. Take four connector pegs and one 4x6 brick. Connect the 4x6 brick to the recently attached connector peg, and on both sides of the 4x6 brick, connect four connector pegs, as illustrated here:

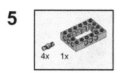

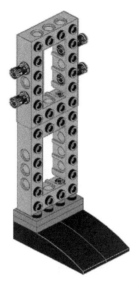

Figure 9.62

6. Take two 6M flex joints and connect these to each side of the connector peg, as illustrated here:

Figure 9.63

7. Take two connector pegs and connect them to the 4x6 brick, as shown here:

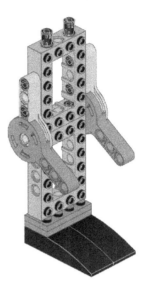

Figure 9.64

8. Now, take one 2x4 blue tile, one 2x4 green tile, one 2x4 red tile, one 2x4 yellow tile, one 2x4 white tile, and one 2x4 brick. Connect the 2x4 brick to the recently attached connector peg and then connect the flat tile to the 4x6 brick, as shown here:

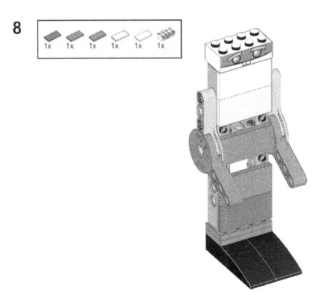

Figure 9.65

9.   Take four 1x1 flat round tiles and one 1x2 brick. Connect the 1x2 brick to the 2x4 brick and connect the 1x1 flat round tile to the 4x6 brick, as shown here:

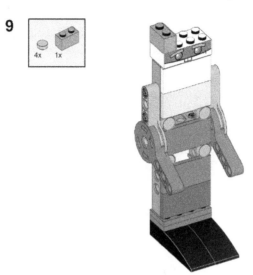

Figure 9.66

10.   Now, take three 1x2 plates and one 2x4 brick with bow. Connect the three 1x2 plates to the 2x4 brick and then connect the 2x4 brick with bow to the 1x2 plate, as illustrated here:

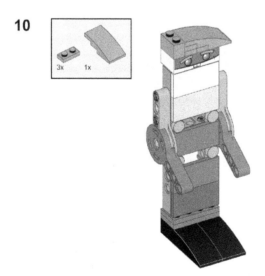

Figure 9.67

Now, let's program this door to open only under certain conditions.

Let's code the door to open under certain conditions

Let's first find out the number of rotations needed to open and close the door completely. Keep the motor power at 30% only. We will need the following number of rotations:

_____ rotations to open the door fully

_____ rotations to close the door fully

Great! Now, do the setup of your automatic door and the color-coded man, as shown in the following screenshot. Identify the right place for the man to stand so that the color sensor can detect the colors accurately. The sensor should be able to sense the color of the hat:

Figure 9.68 – Basic setup of the man and door

Let's first understand the basics of the color sensor and how it works. Your BOOST kit's color sensor can detect six different colors: red, blue, green, yellow, black, and white. It can detect color from a maximum distance of 3 cm. Let's now learn how to code. For sensor-based programming, the first and foremost thing to understand is that the movement of your robot will neither be rotation-based nor seconds-based. So, what will be the logic? Let's understand this with an example. If you want to stop your robot moving when it senses a red color, your basic algorithm will be this:

Move forward until the robot senses a red color. When red is sensed, stop moving.

Based on this, we will write the program. You will be using blocks such as **wait until**, **if**, **if-else**, and **repeat until** from the **Control** pallet and **seeing any color brick** from the BOOST pallet. These blocks can be seen in the following screenshot:

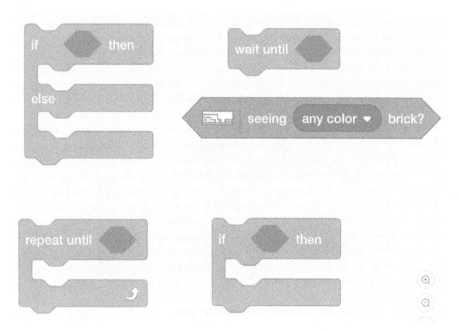

Figure 9.69 – Blocks to use for sensor-based programing

These blocks will help in decision making based on the inputs received from the sensor. **If-else** is a conditional statement for the decisions. **Wait until** is a block which executes the block above it, until the condition inside the **wait until** block is fulfilled. Similarly, **repeat until** block repeats the execution of the blocks inside this repeat block, until the specified condition is met. Use of these blocks will be made more clear in the upcoming activities as well as projects. Let us now move on to write our first sensor based programming.

Activity #1

Write a program to open the door when a blue color is detected. Keep it open for 3 seconds and then close the door. Display a warning message/play a sound file of your choice before closing the door.

The sample code for this activity is given here:

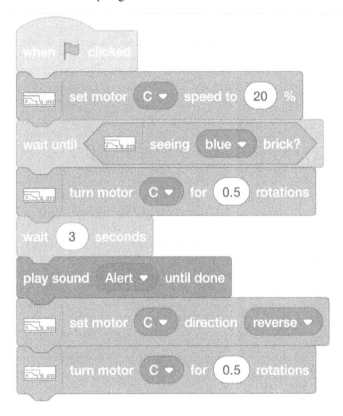

Figure 9.70 – Sample code

Do you remember seeing an **if-then** block in the **Control** pallet? We will be using this block for the next activity.

Activity #2

Open and close the door for different durations for different colors sensed—for example, 2 seconds if blue is detected, 4 seconds if red is detected, and 7 seconds if green is detected.

Basic sample code to perform this activity is illustrated here:

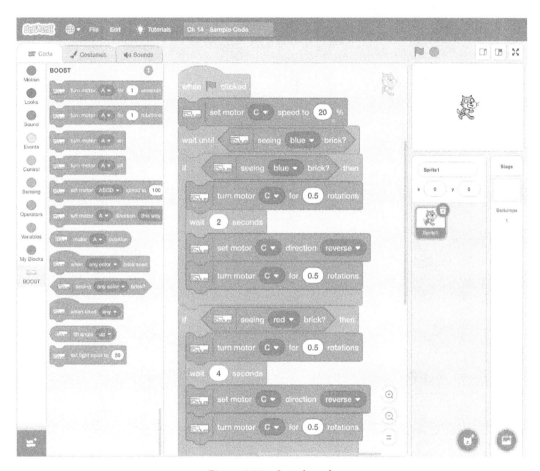

Figure 9.71 – Sample code

Here is a sample code for keeping the door open for 7 seconds when green is detected:

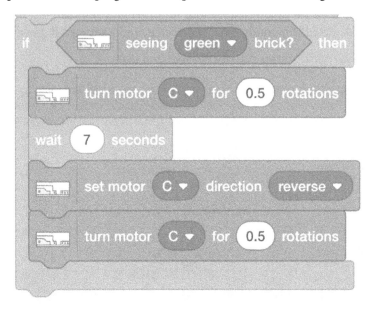

Figure 9.72 – Sample code

You can simply remove the blue hat and change the hat colors accordingly. Refer to the following screenshot:

Figure 9.73 – LEGO bricks for hat colors

You can cascade multiple **if-then** blocks for various conditions that you wish to implement for your robot. Let's now solve the final challenge.

Time for a challenge

Challenge #1

Assume that a door is fixed at an office and the owner wants to allow entry of only specific types of people into the office. So now, program your door to do the following:

1. Open when it senses people with green and blue hats.

2. Keep the door closed for people with red hats. Display **Sorry, entry is restricted** to such people.

You need to use the same man that you built using the BOOST elements. Keep changing the hat color to test it.

Challenge #2

Try to add more colored hats, such as black, white, and yellow hats, using your LEGO elements, and modify the list as follows:

1. Open the door for people with blue, green, and yellow hats.

2. Keep the door closed for people with white, black, and red hats. You can display the same warning message.

You can use more of these `if-else` statements and add to the decision-making capabilities of your BOOST robot for many such exciting projects.

Summary

In this chapter, you worked further with the color sensor and wrote programs for complex tasks with multiple decisions involved. You used an **if-then** block for the first time. You also wrote programs for different actions of the robot for different colors detected. A pulley mechanism was used to open and close the doors. You understood the practical application of automatic doors in this chapter. To further instill these concepts, you will be building and coding a candy dispenser in the next chapter that will dispense candy depending on the color-coded currency detected.

Further reading

Most companies give a **radio-frequency identification** (**RFID**)-based ID card to their employees so that it becomes easy to track their attendance as well as ensure entry of only authorized people into the office. You can find out more about how RFID works at `https://lowrysolutions.com/blog/how-rfid-and-rfid-readers-actually-work/`.

Ready-made RFID tags are available on the market. You can interface them with an Arduino programming board and build your own RFID-enabled items in the future.

10
Building a Candy Dispenser Robot

Candy dispensers are one of the most loved things by kids at places such as amusement parks, malls, airports, and elsewhere. They usually come with a simple dispensing mechanism that gives out candy to kids based on the cents inserted into the dispensers. Today, you will be building a candy dispenser that will give out candy based on the color of bricks detected. This project will further enhance your coding skills with the LEGO BOOST color sensor.

You can see an example of an automatic candy dispensing machine here:

Figure 10.1 – Automatic candy dispensing machine

This chapter will cover the following topics:

- Building a candy dispenser robot
- Let's code the robot to dispense candies based on different colors of LEGO brick detected
- Time for a challenge

Technical requirements

In this chapter, you will need the following:

- LEGO BOOST kit with six AAA batteries, fully charged
- Laptop/desktop with Scratch 3.0 programming installed and an active internet connection
- A diary/notebook with a pencil and eraser

Building a candy dispenser

Let's build a robot by following the building instructions given next. This is what our robot should eventually look like:

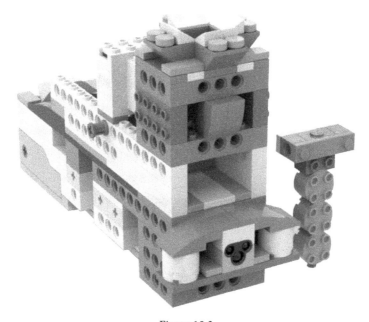

Figure 10.2

Proceed as follows:

1. Take your BOOST Hub and ensure that the batteries are fully charged. The hub is illustrated in the following figure:

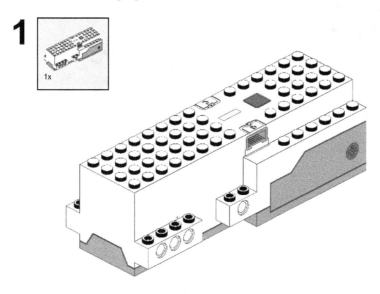

Figure 10.3

2. Take three 1x2 bricks with cross holes and a 1x6 brick and connect them to the BOOST Hub, as shown here:

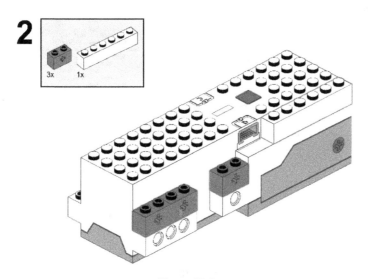

Figure 10.4

3. Take a 1x2 plate and connect it to one of the green bricks placed on the BOOST Hub. Then, take a 1x4 plate and place it onto the other two green-colored bricks, as shown here:

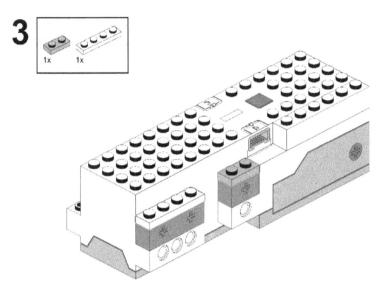

Figure 10.5

4. Now, take a 1x2 brick and connect it to the 1x2 plate, as follows:

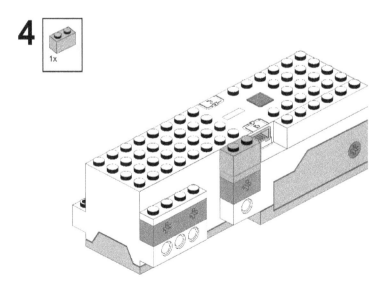

Figure 10.6

5. Then, change the side of the model. Take three 1x2 bricks with cross holes and one 1x6 brick and connect them to the BOOST Hub, as shown here:

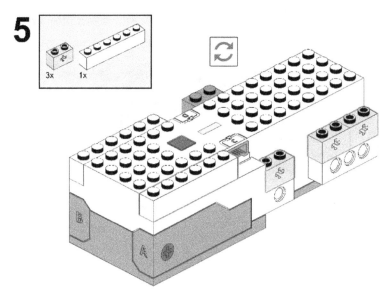

Figure 10.7

6. Take one 1x2 plate and one 1x4 plate. Connect them to the previously attached 1x2 yellow-colored bricks, as shown:

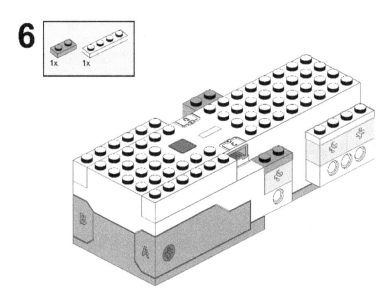

Figure 10.8

7. Take one 1x2 brick and connect it to the 1x2 plate, as follows:

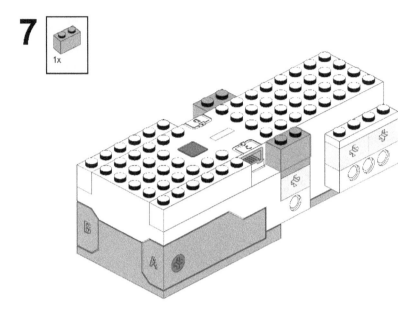

Figure 10.9

8. Take two 1x10 bricks and connect them to both the 1x4 plates, as shown here:

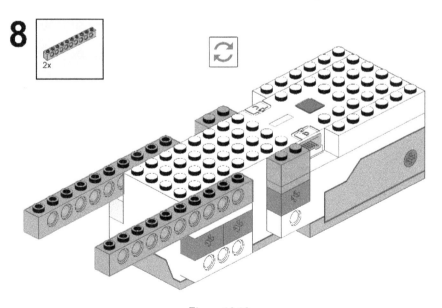

Figure 10.10

9. Take three 4x6 bricks, make a stack of them, and connect this stack below the 1x10 bricks, as illustrated here:

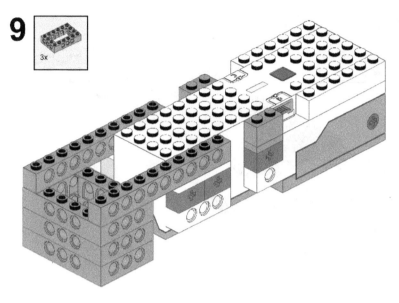

Figure 10.11

10. Take one color sensor from your BOOST kit and attach it to the stack of 4x6 bricks, as shown here:

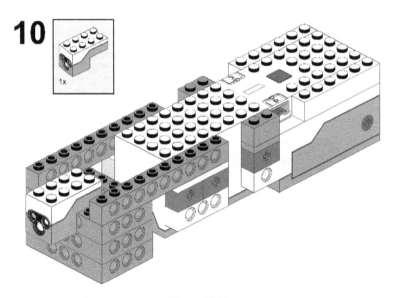

Figure 10.12

11. Take one 6x10 plate and connect it in such a way that it covers the BOOST Hub and the connected bricks, as follows:

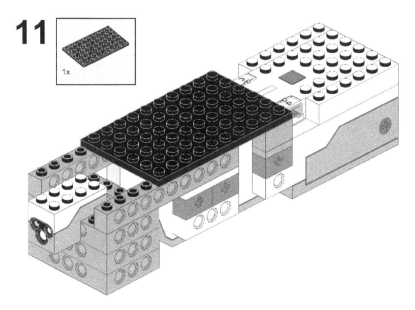

Figure 10.13

12. Now, take two 2x6 plates and connect them to the color sensor and the two 1x10 bricks, as shown here:

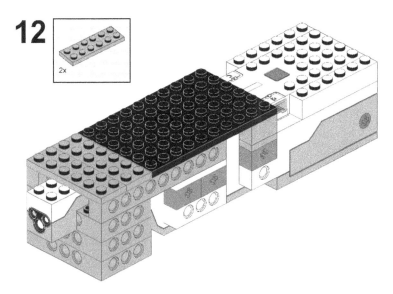

Figure 10.14

13. Take two 1x2 flat tiles and one 2x2 flat tile, then connect them to the BOOST Hub, as shown here:

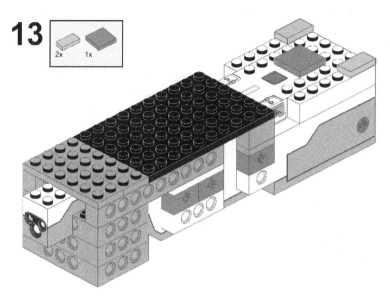

Figure 10.15

14. Now, take two 2x6 bricks with bows and connect them to the BOOST Hub, as follows:

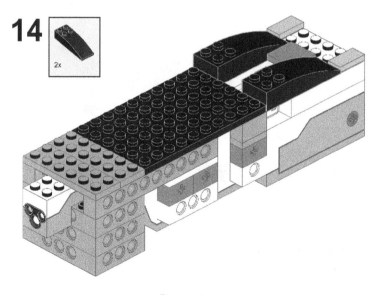

Figure 10.16

15. Take four 2x2 plates, make two stacks with two plates, and connect them to the bricks with bows. Then, take one 4x3 brick with a bow/angle and connect it between the two orange tiles, as shown here:

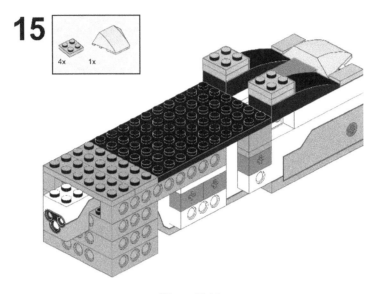

Figure 10.17

16. Now, take two 1x6 plates and connect them to the 1x10 black plate, as follows:

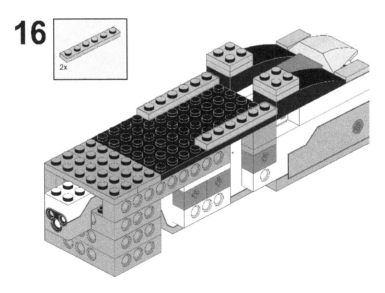

Figure 10.18

17. Take two 1x8 plates and connect them to the black plate and both blue plates, as shown here:

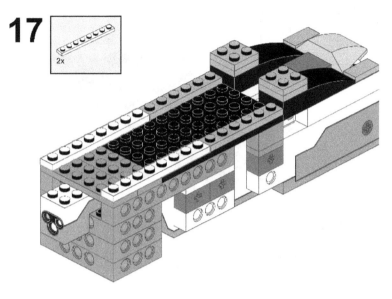

Figure 10.19

18. Then, take two 1x8 flat tiles and connect them beside the orange and white plates, as illustrated here:

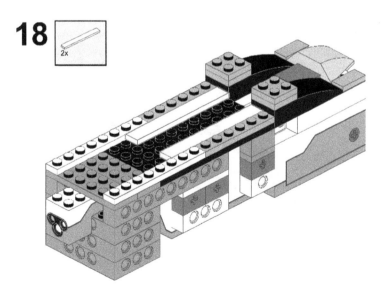

Figure 10.20

19. Then, take two 1x2 flat tiles and two 1x4 flat tiles and connect them to the model, as shown here:

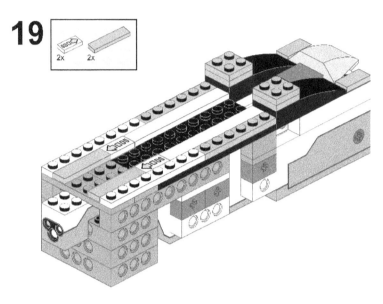

Figure 10.21

20. Take two 1x4 flat tiles and connect them between the two white tiles, then take one 2x4 flat tile and connect it between the two orange tiles, as illustrated here:

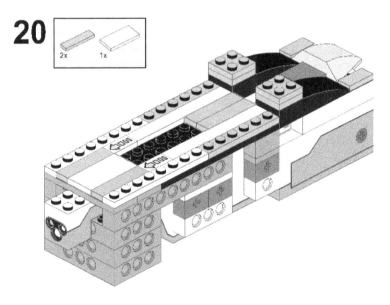

Figure 10.22

21. Take two 1x2 flat tiles and two 1x4 flat tiles and connect them to the central black part, as illustrated here:

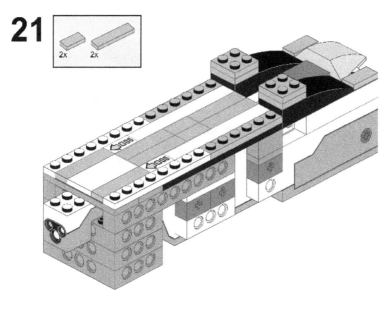

Figure 10.23

22. Take two 1x8 bricks and connect them to both the 1x6 orange plates, as shown here:

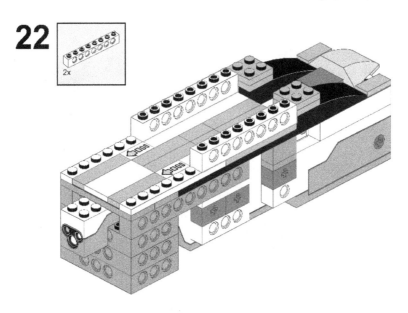

Figure 10.24

23. Take two 1x6 bricks and connect them to the white-colored plates, as shown here:

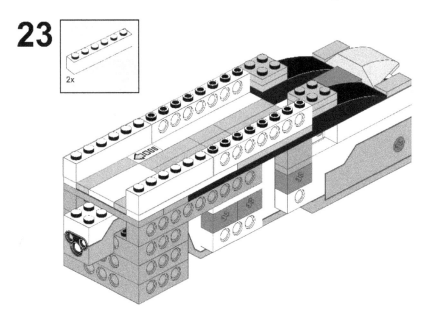

Figure 10.25

24. Take two 1x16 bricks and place them onto the white-colored bricks, as shown here:

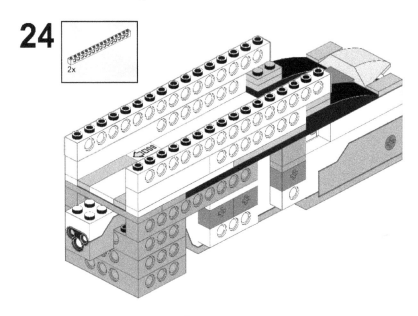

Figure 10.26

25. Take two 2x4 bricks with bows and connect them to both sides of the sensor, as shown here:

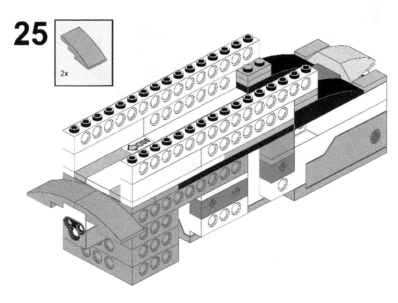

Figure 10.27

26. Then, take two 2x2 round bricks with crosses and connect them under the 2x4 bricks with bows, as shown here:

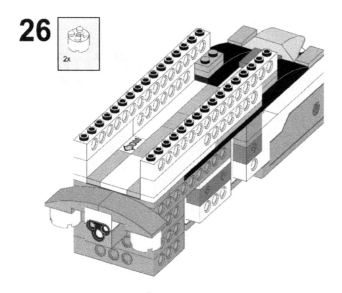

Figure 10.28

27. Take two 2x2 round plates and connect them under the bricks with crosses, as follows:

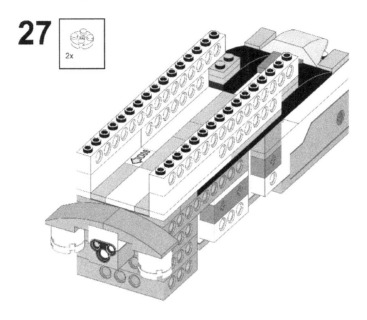

Figure 10.29

28. Now, take one 2x8 plate, connect two 1x2 flat tiles to it, and connect this to the round plates, as shown here:

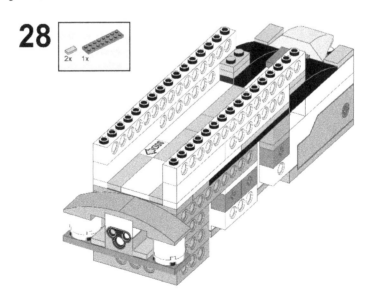

Figure 10.30

29. Take one 8M cross axle with a stop and one Z12 double conical wheel, then attach them to each other and attach the axle to the brick, as shown here:

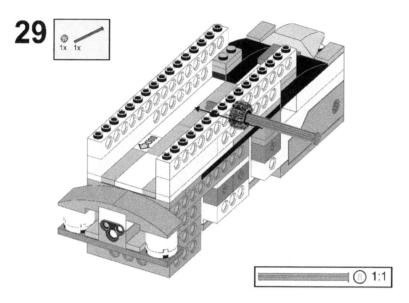

Figure 10.31

30. Then, take one 1M 8T gear wheel, one tube with a double hole, and one bush, then connect them one by one to the axle, as shown here:

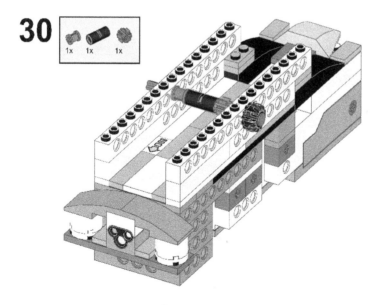

Figure 10.32

31. Now, take one bush and connect it to the axle. Then, take two 1x2 plates and connect them to the brick, as shown here:

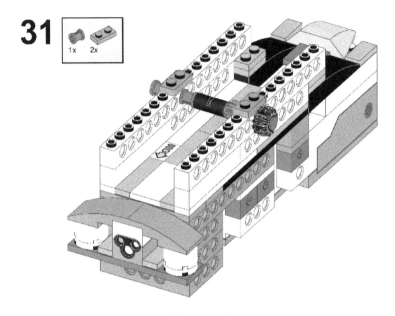

Figure 10.33

32. Now, we are going to make a sliding mechanism. For that, take one 2x4 plate and two 4x4 plates and put them together, as follows:

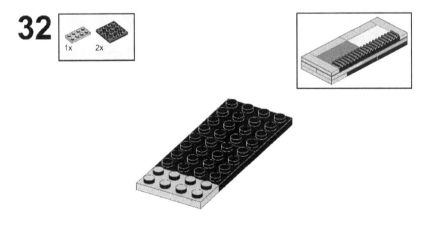

Figure 10.34

33. Take two 2x4 plates and two 2x6 plates, and use them to connect the plates we used before, as follows:

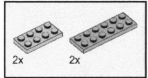

Figure 10.35

34. Then, take two toothed bars and connect them to the plates, as shown here:

Figure 10.36

35. Take one 2x4 flat green tile and one 2x4 flat yellow tile and connect them beside the toothed bar, as shown here:

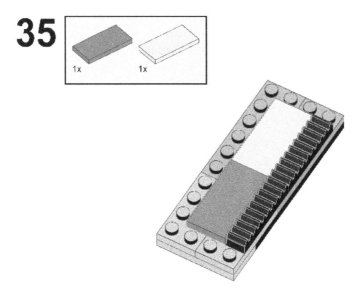

Figure 10.37

36. Then, take four 1x4 flat tiles and cover the remaining area, like this:

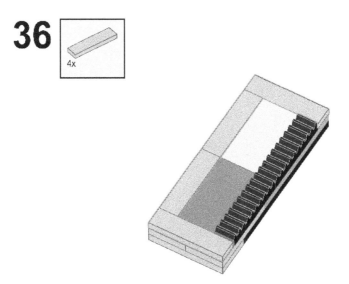

Figure 10.38

37. Now, fix this slider with the original model in such a way that the rack of the slider connects to the 1M 8T gear wheel, as shown here:

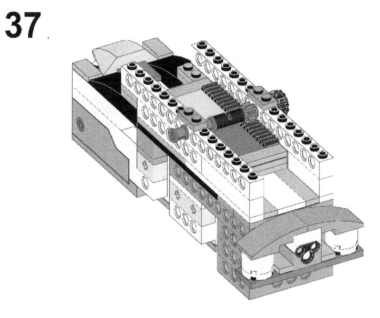

Figure 10.39

38. Take an external motor from your BOOST kit and one 2M cross axle. Connect the cross axle to the motor. Then, fix this motor to the 1x2 blue-colored plates, as shown here:

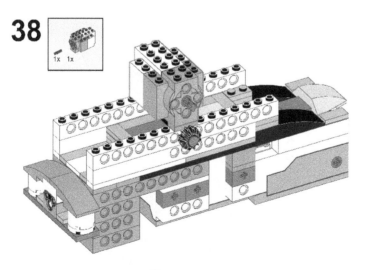

Figure 10.40

39. Now, take one Z36 double conical wheel and connect it to the 2M cross axle, as follows:

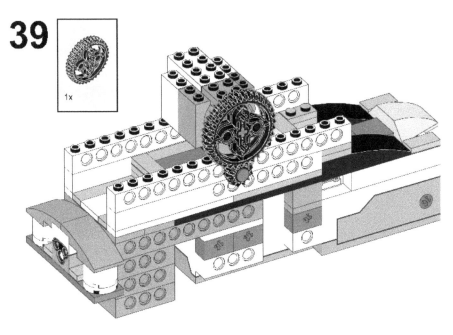

Figure 10.41

40. Now, we will make a trunk in which we can put the candy. For that, let's take a 4x6 brick, as follows:

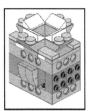

Figure 10.42

41. Now, take six 1x2x1 2/3 bricks with four knobs and fix them onto the 4x6 brick, as follows:

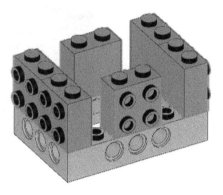

Figure 10.43

42. Then, take one more 4x6 brick and place it onto the red-colored bricks with four knobs. Then, take two 1x4 plates with two knobs and connect them to the 4x6 bricks, as follows:

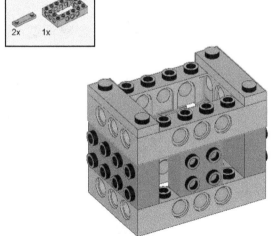

Figure 10.44

43. Take two 1x4 plates and two 1x2 plates and connect them, as shown here:

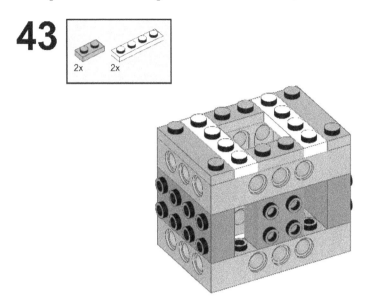

Figure 10.45

44. Take four 1x2 flat tiles and connect them to those plates, as shown here:

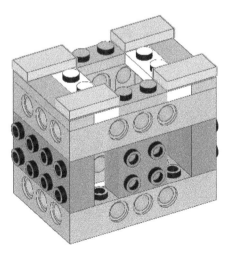

Figure 10.46

45. Take four 2x2 4/5 inverted roof tiles and connect them to the plates, as follows:

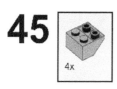

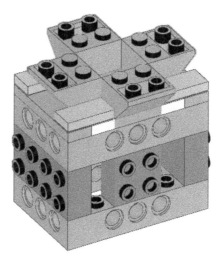

Figure 10.47

46. Take four 1x2 2/3 roof tiles and connect them, as shown here:

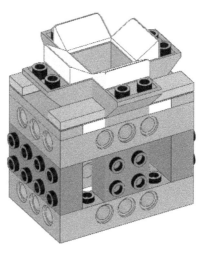

Figure 10.48

47. Take eight 1x1 flat round tiles and connect them behind the roof tile, as shown here:

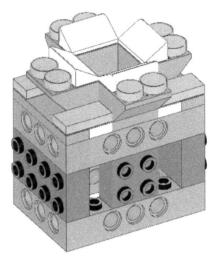

Figure 10.49

48. Take one 1x2 plate and connect it to the red-colored brick, as shown here:

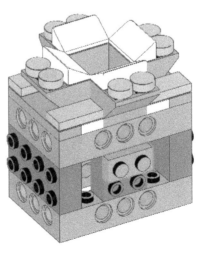

Figure 10.50

49. Then, take two 1x2x2/3 plates with bows and connect them to the 1x2 plate and red-colored brick, as follows:

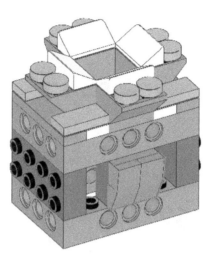

Figure 10.51

50. Now, attach this trunk-like structure to the front side of the main model, as shown here:

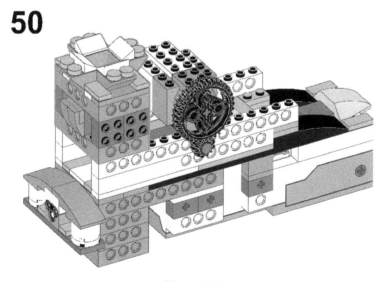

Figure 10.52

51. Now, let's make a candy holder. For that, take one 1x4 plate, as follows:

Figure 10.53

52. Then, take one 1x2 brick with a cross hole and two 1x1 bricks, then connect them to the plate, as follows:

Figure 10.54

53. Take one 1x4 plate with two knobs and connect it to the bricks, as follows:

Figure 10.55

54. Take one 8M stop axle and connect it to the brick with a cross hole, as shown here:

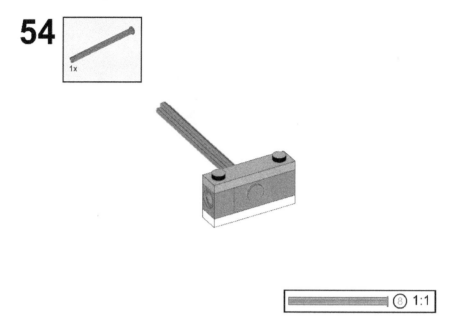

Figure 10.56

55. Then, take one 1x2 brick and connect it to the axle, as follows:

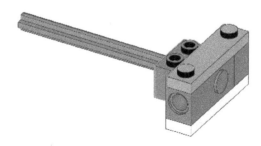

Figure 10.57

56. And here is your candy holder, ready to use:

56

Figure 10.58

57. Take three 2x2 round bricks with holes and use them as candy, as illustrated here:

57

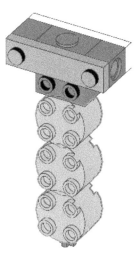

Figure 10.59

Bingo!

Your candy dispenser is ready to use, as we can see here:

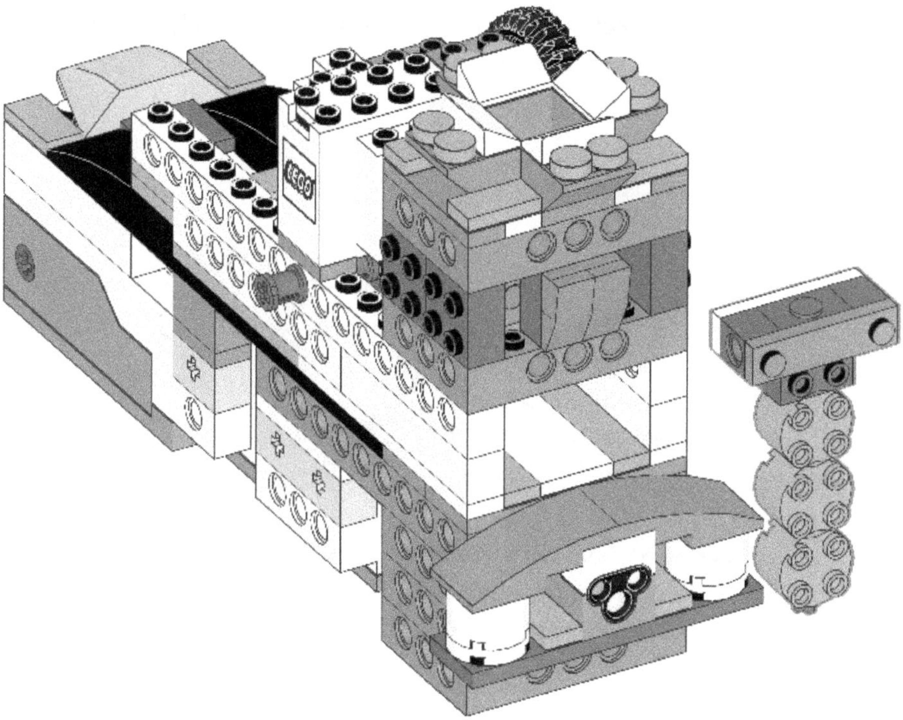

Figure 10.60

Connect this external motor to port C. Let's now code it to dispense candies, based on the color of LEGO brick detected.

Let's code the robot to dispense candies based on different colors of LEGO brick detected

Take three different-colored LEGO bricks—blue, green, and yellow. Now, through trial and error, find out the number of rotations needed by the motor to dispense one piece of candy, and note it here:

Rotations of port C = _____

Activity #1

Write a program to dispense one piece of candy when the color sensor senses a yellow brick. Display **Yellow Brick Detected** before dispensing. Once the candy is dispensed, display **One Candy Dispensed**, as illustrated in the following screenshot:

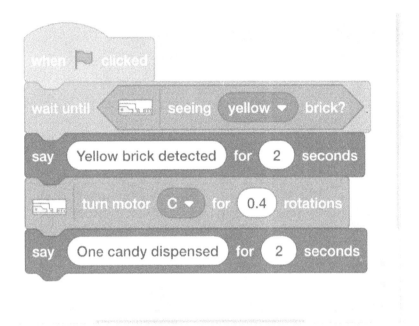

Figure 10.61 – Sample code

Let's move on to the next activity, to add different dispensing actions to the robot for different colors detected.

Activity #2

Write a program to dispense two pieces of candy when the color sensor senses a blue brick. Display **Blue Brick Detected** before dispensing. Once the candy is dispensed, display **Two Candies Dispensed**.

Activity #3

Write a program to dispense three pieces of candy when the color sensor senses a green brick. Display **Green Brick Detected** before dispensing. Once the candy is dispensed, display **Three Candies Dispensed**.

Time for a challenge

Challenge #1

Write one single program for the candy dispenser to dispense, as follows:

1. One piece of candy when yellow is detected
2. Two pieces of candy when blue is detected
3. Three pieces of candy when green is detected

Try to add unique features such as display and sound blocks to make this activity more interactive and user-friendly. You will have to use `if-else` conditions to set up the logic for this activity. You can consider displaying money by the application of variables and see how much a single user can spend.

Summary

In this chapter, you learned how a candy dispenser works. Instead of cents, you used the property of a color sensor to dispense different amounts of candy based on distinct colors detected. You also learned how to build a simple dispensing mechanism using gears and other LEGO elements that can be effectively used for other applications as well. In the next chapter, you will be building an automatic color sorter using a conveyor belt. Again, this is an industrial application used widely for various tasks related to sorting. You will enhance your coding skills using a color sensor and add multiple conditions for the robot to check and act upon.

Further reading

PEZ candy dispensers were extremely popular in the 1990s. Read more about them at `https://www.mentalfloss.com/article/70010/10-indispensable-facts-about-pez`.

You can build such dispensers for your home for various purposes, not only for candy, by applying your construction skills.

11
Building a Color-Sorter Conveyor Belt

With the introduction of the Industrial Revolution 4.0, most industries are now turning toward automation for their production. Huge assembly lines are deployed, and multiple robots are lined up around these assembly lines to do various tasks one after another, automatically. A whole car body is prepared automatically with the help of robots today! Just as we need robots for such processes, we also need robots for sorting various things, ranging from faulty products to different products, into different trays. In this chapter, you will be building one such assembly line that sorts different-colored LEGO bricks into different trays.

You can see an example representation of an assembly line here:

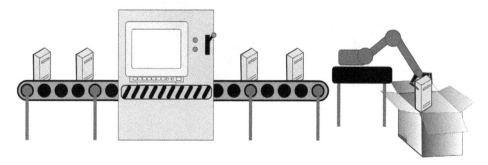

Figure 11.1 – Assembly-line representation

In this chapter, you will do the following:

- Building a color-sorter conveyor belt
- Let's code the robot to sort different colored LEGO bricks
- Time for a challenge

Technical requirements

In this chapter, you will need the following:

- LEGO BOOST kit with six AAA batteries, fully charged
- Laptop/desktop with Scratch 3.0 programming installed and an active internet connection
- A diary/notebook with a pencil and eraser

Building a color-sorter conveyor belt robot

Let's build a robot like the one shown in the following figure:

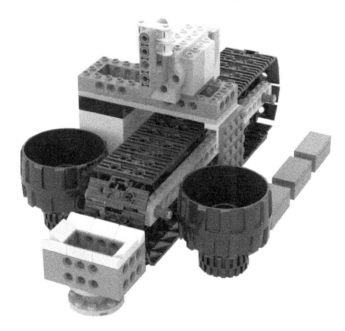

Figure 11.2

Follow these given instructions to build the robot:

1. Take your BOOST Hub and ensure that the batteries are fully charged. The Hub is illustrated in the following figure:

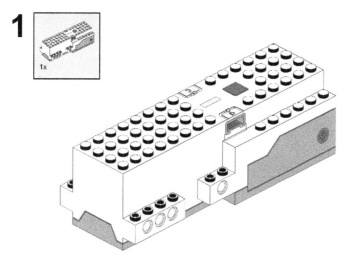

Figure 11.3

2. Take four 1x6 bricks and connect them to the back side of the BOOST Hub, as follows:

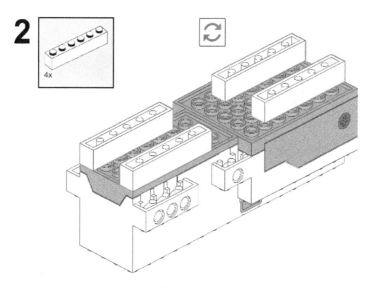

Figure 11.4

3. Take four 1x3 bricks and connect them to all four 1x6 bricks, then take one 4x6 brick and place it so that it will connect to all four 1x6 bricks, as shown here:

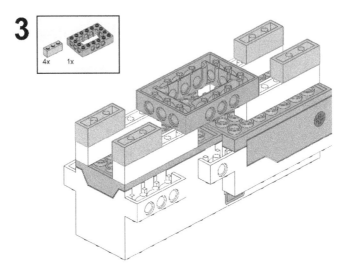

Figure 11.5

4. Now, take two 6x10 plates and connect them to two 1x3 bricks and the 4x6 brick, as shown here:

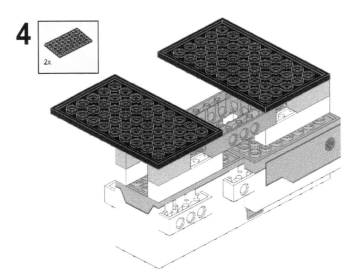

Figure 11.6

5. Now, turn the model frontside, take two 2x4 bricks, and place them on the BOOST Hub, as follows:

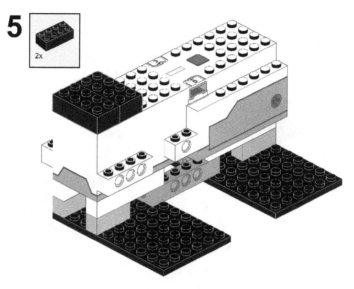

Figure 11.7

6. Take three 3M connector pegs with friction and connect them to the BOOST Hub, as shown here:

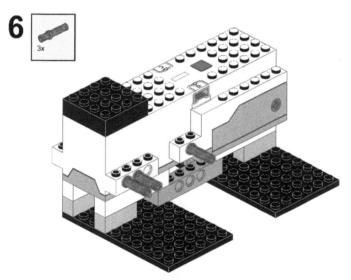

Figure 11.8

7. Then, take two 1x10 bricks and connect one of them to two pegs and another one to a single peg, as shown here:

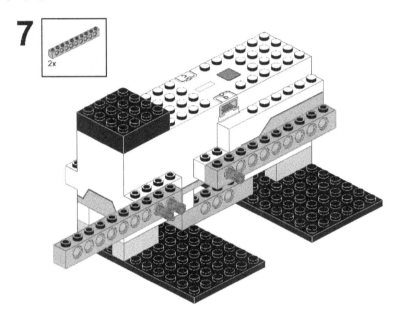

Figure 11.9

8. Now, take one 1x6 plate and one 1x8 brick and connect this brick to the plate, so that half of the plate will be covered. This is illustrated here:

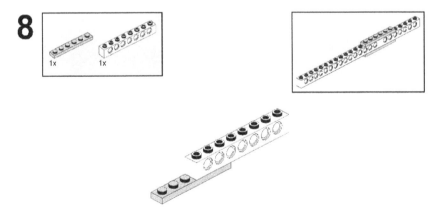

Figure 11.10

9. Take one 1x16 brick and connect it to the 1x6 orange plate, where you just connected the 1x8 brick. Now, take one 1x6 plate and connect it to the top of these two bricks, as shown in the following figure. This will hold the bricks firmly. Again, take one 1x6 plate and connect it to both the bricks:

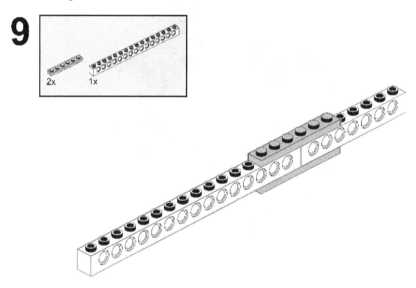

Figure 11.11

10. Then, connect this structure to the 3M connector pegs, as shown here:

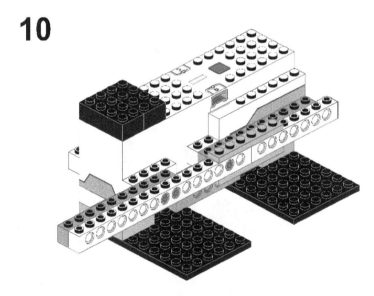

Figure 11.12

11. Take two 7M cross axles and connect them to the brick. Then, take one bush and connect it to stop one of those axles. The result should look like this:

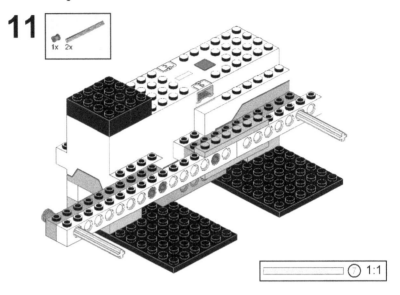

Figure 11.13

12. Now, take two sprockets and connect them to both the axles. Then, take one 4x4 plate and connect it to the 6x10 plate, as shown here:

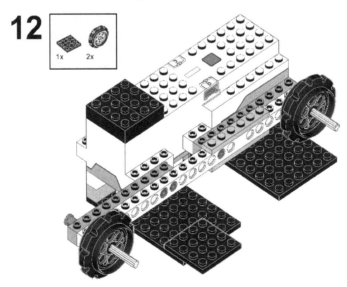

Figure 11.14

13. Take 10 track elements and connect them all, as shown here:

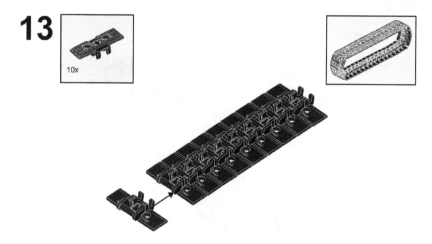

Figure 11.15 –

14. Take 30 such track elements and connect them with each other. Once done, connect this track with the track you built in the previous step. We will now get a conveyor belt as per our requirement, as follows:

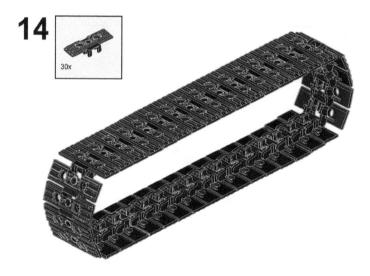

Figure 11.16

15. Now, connect this round-shaped track with those two sprockets. Then, take one 2x4 plate and connect it under the 4x4 plate, as follows:

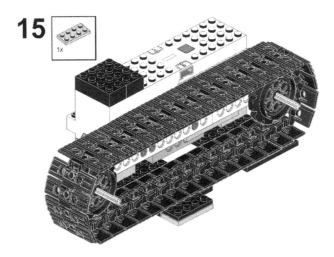

Figure 11.17 –

16. Now, take one 1x6 plate and one 1x8 brick and connect this brick to the plate that we have used (so that half of the plate will be covered). This is illustrated here:

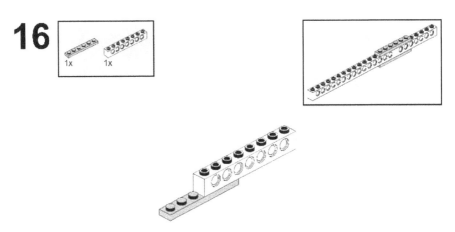

Figure 11.18

17. Then, take one 1x16 brick and connect it to the remaining half of the plate. Again, take one 1x6 plate and connect it to both the bricks, as follows:

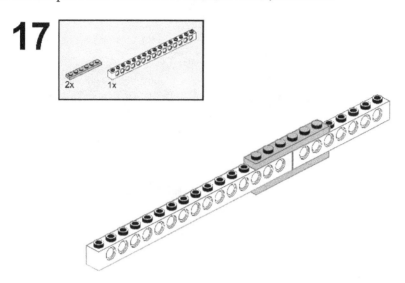

Figure 11.19

18. Now, connect this structure to both the axles, as follows:

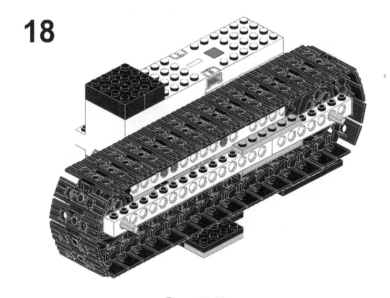

Figure 11.20

19. Take two bushes and connect them to both the axles, then take two 1x2/2x2 angular plates and connect them to the 4x4 plate, as shown here:

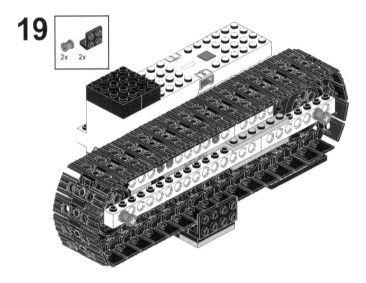

Figure 11.21

20. Then, take two 2x8 plates and connect them to the angular plates, as follows:

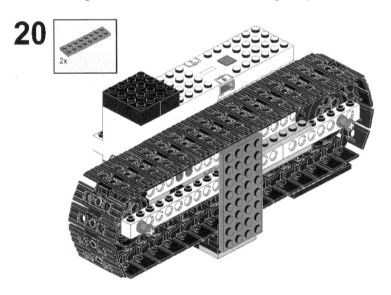

Figure 11.22

21. Take two 1x2/2x2 angular plates and connect them to both the 2x8 plates, as shown here:

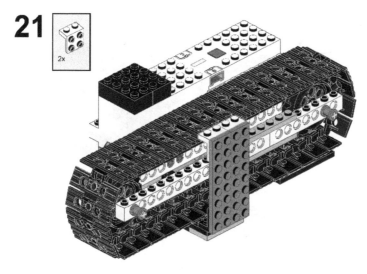

Figure 11.23

22. Then, take two 2x12 plates and connect them to the 2x4 bricks and angular plates, as follows:

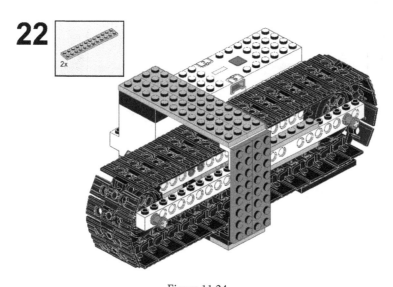

Figure 11.24

23. Take two 4x6 bricks and connect them to the 2x12 plates. Then, take two 1x4 plates and connect them, as shown here:

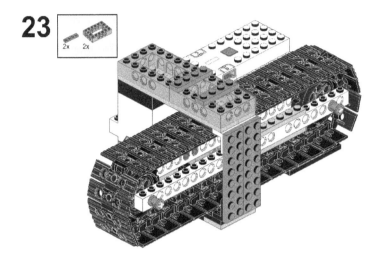

Figure 11.25

24. Take one 5M beam and three 3M connector pegs and connect all of them to the 5M beam (two with the first two holes and one with the fourth hole of the beam), as follows:

 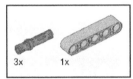 

Figure 11.26

25. Take two double-cross blocks and connect them to the first two 3M pegs, then take one 4M stop axle followed by one 1x2 beam with a cross and hole, connect the axle to the last hole of the 5M beam, and connect the 1x2 beam to the cross and hole, connecting it with the remaining 3M peg and axle. The process is illustrated here:

Figure 11.27

26. Again, take one 5M beam and connect it as shown, then take two 2M axles and connect them to both the cross-hole blocks, as shown here:

Figure 11.28

27. Now, take an external motor from your BOOST kit and connect the axle of the model to it, as follows:

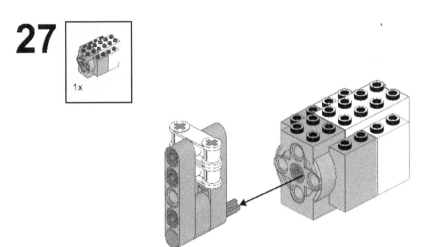

Figure 11.29

28. Now, place the motor on the 1x4 plates, as follows:

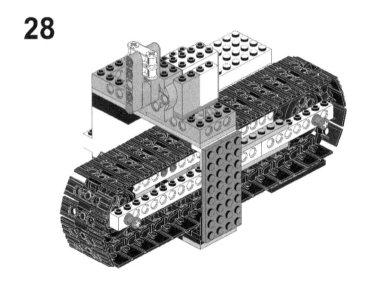

Figure 11.30

29. Take one 1x2 plate and one 1x2 1/2 angular plate, then connect them to each other, as shown here:

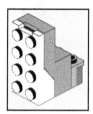

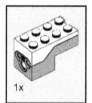

Figure 11.31

30. Take a color sensor from your LEGO BOOST kit and connect this to the orange-colored angular plate, as shown here:

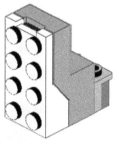

Figure 11.32

31. Now, connect this structure to the motor using the angular beam, as shown here:

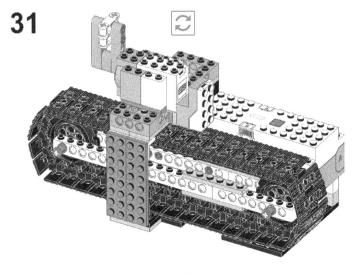

Figure 11.33

32. Now, let's make a bowl to collect the separated boxes. For that, first take one wheel and one brick with a cross and connect it to the hole of the wheel, as follows:

Figure 11.34

33. Then, take one 3M cross axle and connect it to the brick with a cross, as follows:

Figure 11.35

34. Then, take one normal tire and a rim wide with a cross, connect them to each other, and then connect them to the 3M axle, as follows:

Figure 11.36

35. Now, place this bowl beside the circular track to collect the boxes, as follows:

35

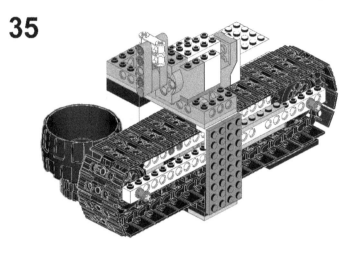

Figure 11.37

36. Now, make one more bowl by following the same steps shown previously. Your bowl should look like this:

36

Figure 11.38

37. Place this second bowl on the other side of the circular track, as illustrated here:

37

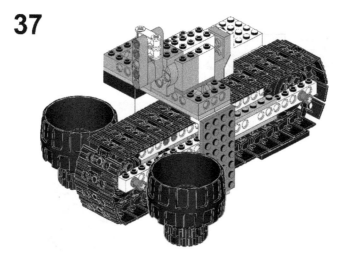

Figure 11.39

38. Now, we are going to make one more bowl to collect the boxes. For that, let's start by taking a 4x4 round-shaped plate and one 2x2 round brick with a hole and connect them, as shown here:

38

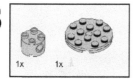

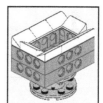

Figure 11.40

39. Take one more 4x4 round-shaped plate and connect it to the 2x2 round brick with a hole, as follows:

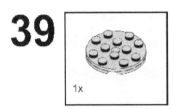

Figure 11.41

40. Now, take two 4x6 bricks and connect them to the 4x4 round-shaped plate, as shown here:

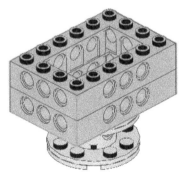

Figure 11.42

41. Then, take eight 1x2x2/3 roof tiles and connect them to the 4x6 brick, as shown here:

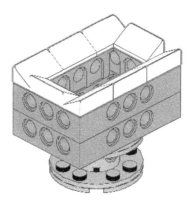

Figure 11.43

42. Now, place this front tray (made of blue LEGO bricks) as shown in the following figure:

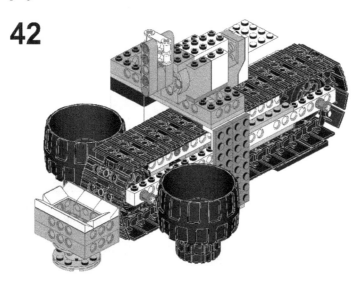

Figure 11.44

Now, let's make colorful boxes to sort in different buckets.

43. Let's start by making a green box. For that, take two 2x2 bricks and place them side by side, as follows:

Figure 11.45

44. Then, take a green flat tile and place it on both the bricks, as follows:

Figure 11.46

45. For the blue box, again take two 2x2 bricks and place them back to back, as follows:

Figure 11.47

46. Take a 2x4 blue-colored flat tile and place it on both the bricks, as follows:

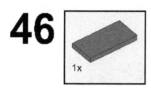

Figure 11.48

47. Now, for the red box, take one 2x4 red-colored brick and one 2x4 red flat tile and connect them to each other, as follows:

Figure 11.49

And you are now done with your color sorter!

Check if your model is working fine and that the connections are sturdy as depicted in the figure:

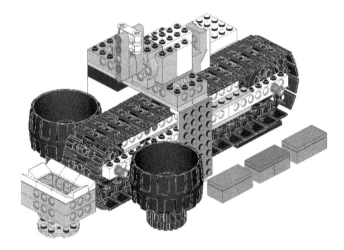

Figure 11.50

Well done! Let's now switch to the coding section and solve various tasks.

Let's code the robot to sort different colored LEGO bricks

Let's now sort the LEGO bricks with this color sorting robot. First, find out the number of rotations needed by the external motor to do the following:

1. Knock LEGO brick to the left tray:_____ rotations.

2. Knock LEGO brick to the right tray:_____ rotations.

3. Allow LEGO brick to pass through at the blue tray:_____ rotations.

Ensure that the hitter comes back to the original position after knocking the LEGO brick. The hitter is connected to port C and the conveyor belt is connected to port A.

Activity #1

Place a red LEGO brick on the belt. Now, move the belt until the color sensor senses this brick. Display the brick color on the screen. Now, move it further until it reaches a proper knocking position. Knock it to the left tray.

The sample code for this task is given here:

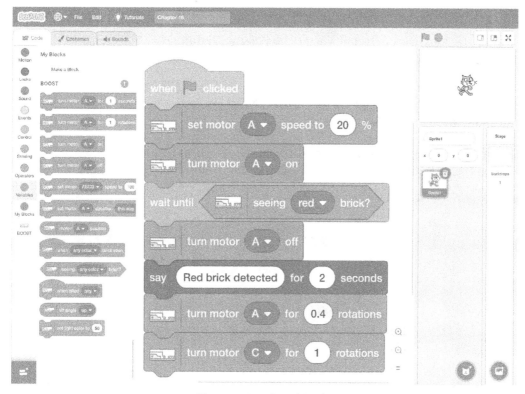

Figure 11.51 – Sample code

Let's repeat the same process for the other colors as well.

Activity #2

Place a blue LEGO brick on the belt. Now, move the belt until the color sensor senses this brick. Display the brick color on the screen. Now, move it further until it reaches a proper knocking position. Knock it to the right tray.

Activity #3

Place a green LEGO brick on the belt. Now, move the belt until the color sensor senses this brick. Display the brick color on the screen. Now, move it further until it reaches a proper knocking position. Let it pass through and drop down in the front tray (made with blue LEGO bricks).

Activity #4

Now, let's just take two bricks—red and blue. Let's now code the sorter to do the following:

1. Display **Red brick detected** when red is detected and **Blue brick detected** when blue is detected.

2. Knock the red brick to the left tray and the blue brick to the right tray.

3. If the brick is neither red nor blue, display **Wrong brick on the belt! Pls check" as in the following screenshot**.

Both these things must work continuously, regardless of the order in which you place the bricks on the belt. Make use of the **if else** block wisely. Repeat this task five times.

Here is some sample code to show how to do this:

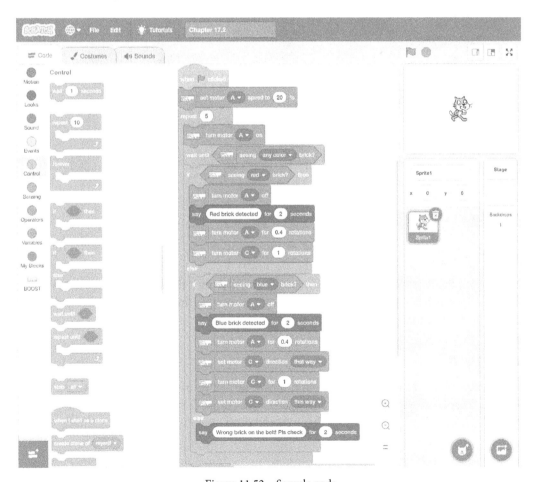

Figure 11.52 – Sample code

You must place the LEGO bricks one after the other on the belt. The second brick must be placed on the belt only when the first brick is sorted and knocked down by the hitter to its respective tray. One important point to note is that the number of rotations for the belt may vary slightly, to transfer the brick from its color sensor stop position to a knocking position. Find this out yourself and input the correct value. Let's now move on to the challenge section.

Time for a challenge

Challenge #1

Just as with *Activity #4*, can you add a green brick and do the respective programming as follows?

1. Red brick to be knocked to the left tray
2. Blue brick to be knocked to the right tray
3. Green brick to be knocked to the front tray (made with blue LEGO bricks)

You must display on the screen the respective color name that is sensed by the sensor. The program must be such that no matter the order in which you place the bricks on the belt, it should be able to sort them out. Make sure to place one brick after another.

Summary

In this chapter, you learned about the practical application of color sorters in an industrial setting. You learned how to use `if-else` conditions effectively and repetitively. You were able to program your robot to make decisions from the multiple conditions it was offered. In the next chapter, you will be building a racing car with an internal steering system that will conquer various racetracks at different speeds and of different complexities.

Further reading

Here are some interesting topics that can be explored:

- See how robotic arms, sorting robots, and other robots build up a smart warehouse/factory at `https://www.youtube.com/watch?v=IMPbKVb8y8s&ab_channel=TechVision`.

- Sorting robots are extensively used in logistics industries, whereby robots scan the barcode on a parcel and sort them based on the cities/pin codes for dispatch. You can see more about this here: `https://www.youtube.com/watch?v=ezym16z1NPg&ab_channel=TheRobotReport`.

12
Building a BOOST Racing Car

So far, whatever models you have built using your BOOST kit have had a simple two-wheel drive feature. In this chapter, you will be building a steering-controlled car with gears. You might have observed that your parents always maneuver their car using the car's steering wheel. Whenever the steering wheel is moved, the wheels of the car move in the respective direction. This helps the car maneuver. To build such cars using LEGO elements, we will be using gears. For a car, it is important to have a stopping position in both directions to ensure our turns are controlled. Let's start building this car:

Figure 12.1 – Racing car

In this chapter, we will cover the following topics:

- Building the racing car
- Let's code the robot to run on different racetracks
- Time for a challenge

Technical requirements

In this chapter, you will need the following:

- A LEGO BOOST kit with 6 AAA batteries, fully charged
- A laptop/desktop with the Scratch 3.0 programming language installed and an active internet connection
- A diary/notebook, along with a pencil and eraser

Building the racing car

In this chapter, we will be building the following racing car:

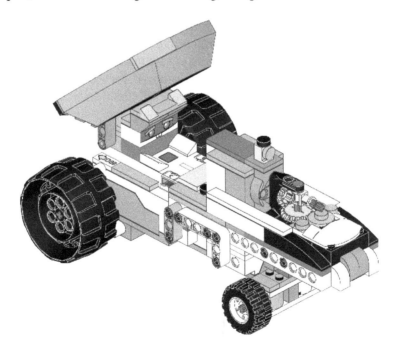

Figure 12.2

Follow these steps to build this racing car robot:

1. Take your BOOST Hub and ensure that its batteries are fully charged:

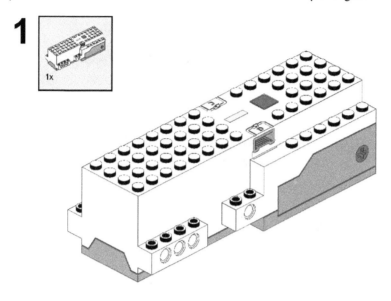

Figure 12.3

2. Take one 4x6 brick and connect it to the back of the hub, as shown in the following figure:

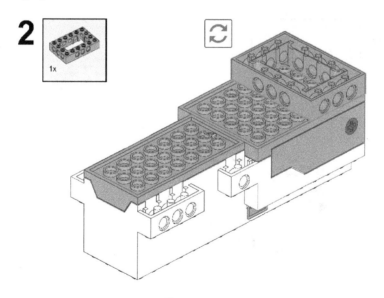

Figure 12.4

3. Take six connector pegs and connect three pegs to each side of the 4x6 brick, as shown here:

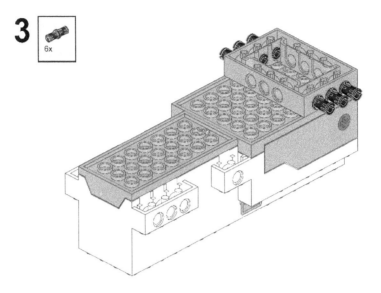

Figure 12.5

4. Now, flip the model, take one 1x2 flat tile and one 1x4 flat tile, and connect both to the side beam of the BOOST Hub, as shown in the following figure. Now, take one 3x5 angular beam and connect it to all three connector pegs on that side:

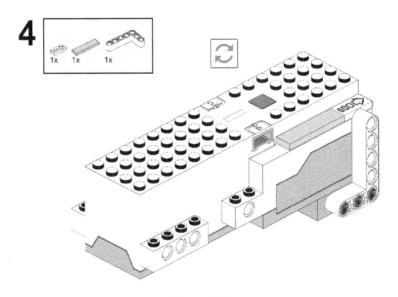

Figure 12.6

5. Now, take one 5M axle with a 1M stop and one ½ bush. First, remove the angular beam, connect the top part of the axle to motor B of the BOOST Hub, connect that to the angular beam, and then connect the ½ bush to the axle:

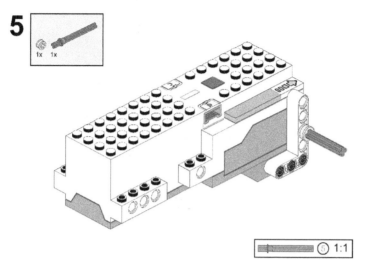

Figure 12.7

6. Now, rotate the hub so that you can see motor A. Again, take one 1x2 flat tile and one 1x4 flat tile and connect both to the side beam of the BOOST Hub.

    Now, take one 3x5 angular beam and connect it to all three connector pegs on that side:

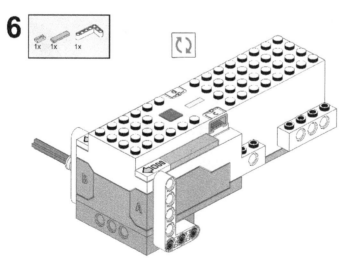

Figure 12.8

7. Now, take one 5M axle with a 1M stop and one ½ bush. Once again, to fix the axle, remove the angular beam and connect the top part of the axle to motor A of the BOOST Hub. Then, connect the angular beam once more and connect the ½ bush to the axle:

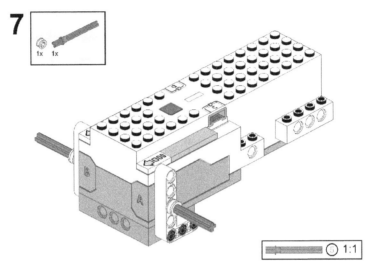

Figure 12.9

8. Take one wheel and one 2x2 round plate and connect the plate to the center part of the wheel.

Take one more 2x2 round plate and connect it to another wheel:

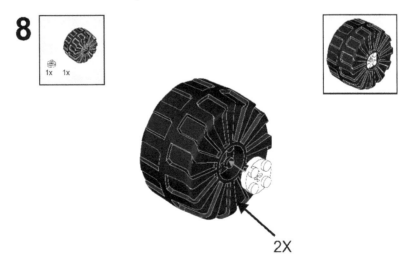

Figure 12.10

9. Now, connect one of those wheels to the axle, which is connected to motor A, as shown in the following figure:

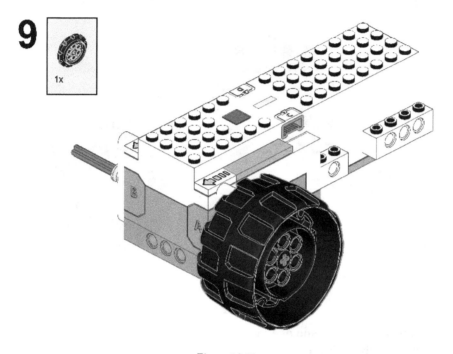

Figure 12.11

10. Now, take another wheel and round plate, which you have already connected:

Figure 12.12

11. Connect this wheel to the axle that is connected to motor B:

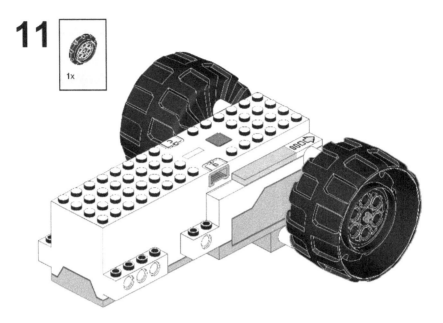

Figure 12.13

12. Then, take one 1x2 plate and one 1x4 plate and connect them to the side bricks of the BOOST Hub, as shown here:

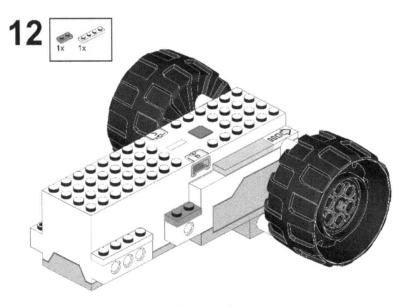

Figure 12.14

13. Take one 1x8 plate and connect it to both plates that you connected in the preceding step. Then, take one 1x16 brick and connect it to the 1x8 plate:

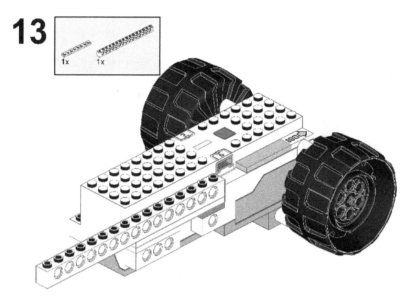

Figure 12.15

14. Now, take five connector pegs. Connect three of them to the 1x16 brick and the remaining two to the side bricks of the BOOST Hub, as shown here:

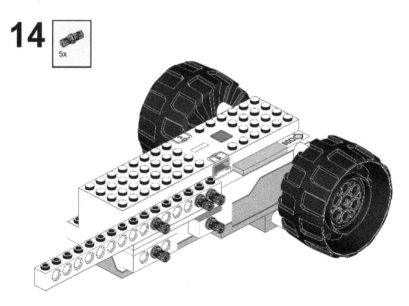

Figure 12.16

15. Now, flip the model so that you can see motor A. Take one 1x2 plate and one 1x4 plate and connect them to the side bricks present on that side:

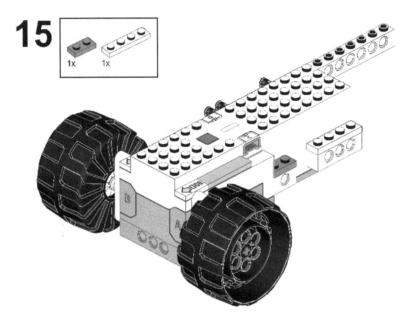

Figure 12.17

16. Again, take one 1x8 plate and connect it to both plates that you connected in the preceding step. Then, take one 1x16 brick and connect it to the 1x8 plate:

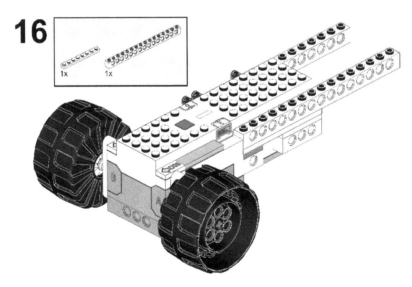

Figure 12.18

17. Now, take five connector pegs. Connect three of them to the 1x16 brick and the remaining two to the side bricks of the BOOST Hub, as shown in the following figure:

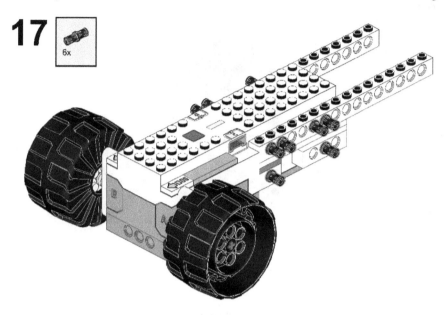

Figure 12.19

18. Now, take one 3M beam and connect it vertically to the connector pegs. Then, take one 3x5 angular beam and connect it as shown in the following figure:

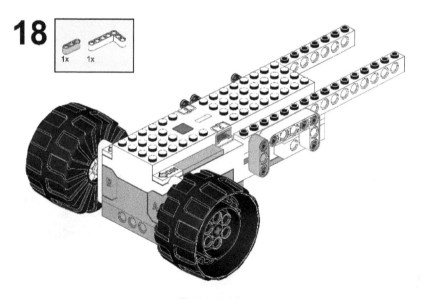

Figure 12.20

19. Do the same thing on the opposite side of the model:

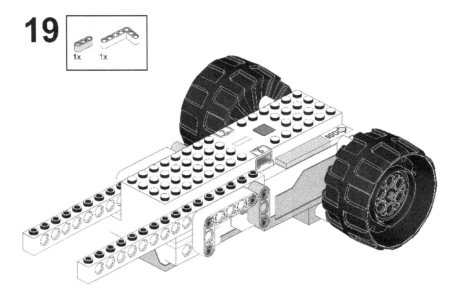

Figure 12.21

20. Then, take one 2x4 brick with a bow and two 2x4 plates. Connect the brick to the BOOST Hub and then connect both plates to that brick, as shown in the following figure:

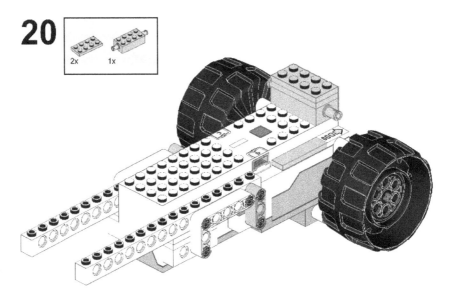

Figure 12.22

21. Now, take one more 2x4 brick with a bow and connect it to the 2x4 plate. Then, take one red 2x4 brick and one 2x4 brick with a face on it. Connect them to the brick with a bow, and then to the BOOST Hub, as shown in the following figure:

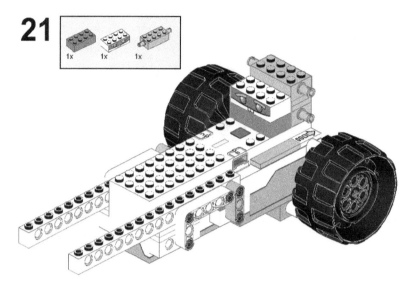

Figure 12.23

22. Take two 1x2 flat tiles and connect them to the brick with a bow, as shown in the following figure. Then, take two 5M beams and connect them to both sides of the brick with a bow:

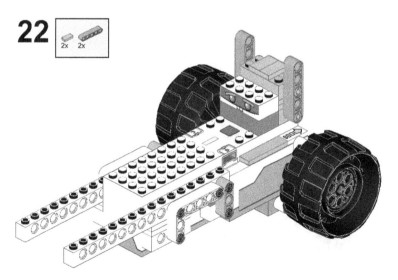

Figure 12.24

23. Now, take two frictional snaps with cross holes and connect them to the first hole of both 5M beams. Then, take one 1x4 flat tile and connect it to the designed brick, as shown in the following figure:

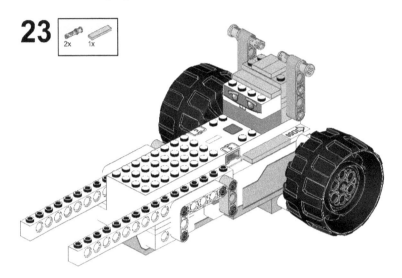

Figure 12.25

24. Take two 2x2 inverted roof tiles and connect them to the designed brick, as shown in the following figure. Then, take two 1x2 bricks and connect them to both frictional snaps:

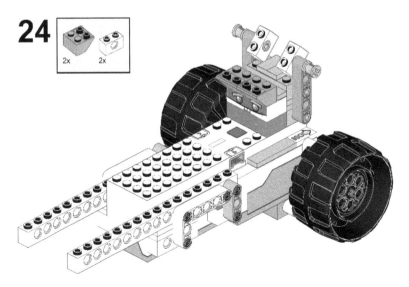

Figure 12.26

25. Take one 1x2 flat tile and one 1x4 flat tile and connect them to the inverted roof tile, as shown in the following figure:

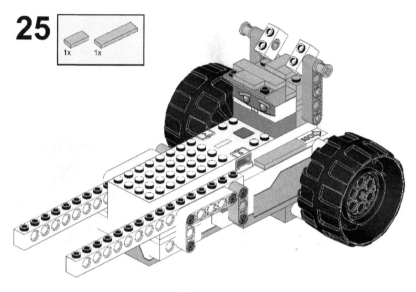

Figure 12.27

26. Take two pink 1x1x2/3 roof tiles and two white 1x2x2/3 roof tiles. Connect the pink roof tiles to both sides of the 1x2 flat tile on top of the inverted roof tile. Then, connect the white roof tiles to both sides of the start button of the BOOST Hub, as shown in the following figure:

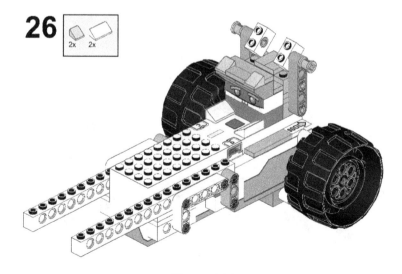

Figure 12.28

27. Take two 1x2 plates and two 1x6 plates and connect them to both 1x16 bricks, as shown in the following figure:

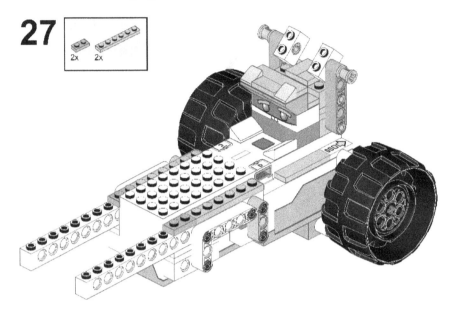

Figure 12.29

28. Take four connector pegs and connect two pegs to each of the 1x16 bricks, as shown in the following figure:

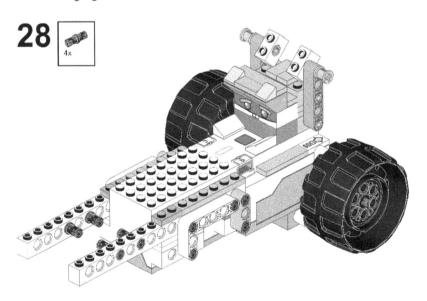

Figure 12.30

29. Now, take one 4x6 brick and connect it to those four connector pegs, between the two 1x16 bricks:

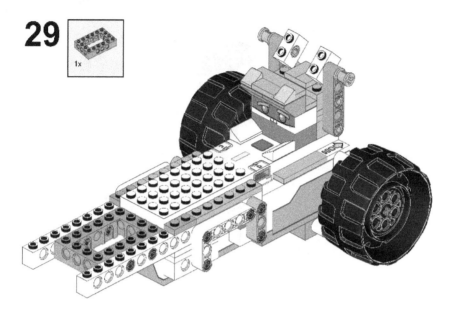

Figure 12.31

30. Take two 2x8 plates and connect them so that they join both 1x16 bricks to the 4x6 brick, as shown in the following figure:

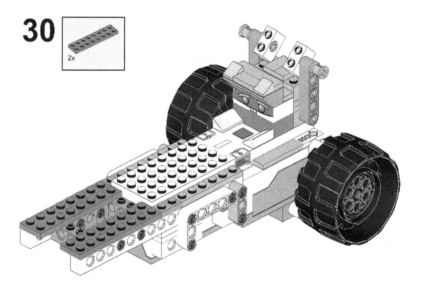

Figure 12.32

31. Now, flip the model back over. Then, take two 1x2 plates and two 2x4 plates and connect them to the back of the 4x6 brick, as shown in the following figure:

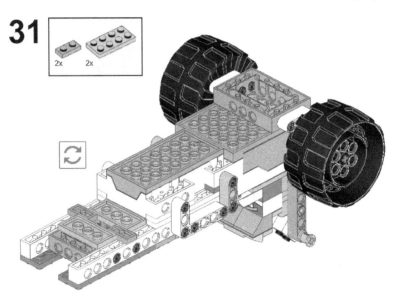

Figure 12.33

32. Take two 1x2 bricks with crosses and one 1M beam with two axles. Connect both bricks to both sides of the 1M beam with axles:

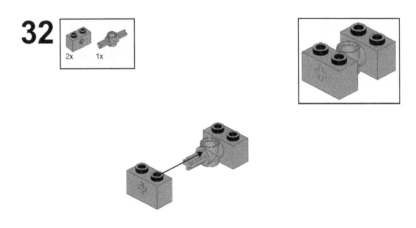

Figure 12.34

33. Now, flip the model on its front and connect this structure to the 2x4 orange-colored plates, as shown in the following figure:

Figure 12.35

34. Now, take two 1x4 flat tiles and connect them to both orange 1x6 plates. Then, take one 2x4 flat tile and connect it to the BOOST Hub, as shown in the following figure:

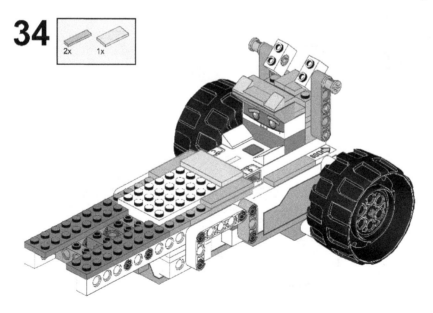

Figure 12.36

35. Take one 2x2 coupling plate and connect it to the red 1x2 brick with a cross. Then, take two orange 1x1 round flat tiles and two black 1x1 round flat tiles and connect them to the BOOST Hub, as shown in the following figure:

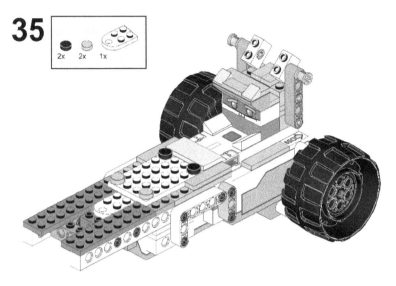

Figure 12.37

36. Take one 2x2 plate and connect it to the coupling plate. Then, take one 2x3 plate and connect it so that it's between the 1x2 brick and the 4x6 brick, as shown in the following figure:

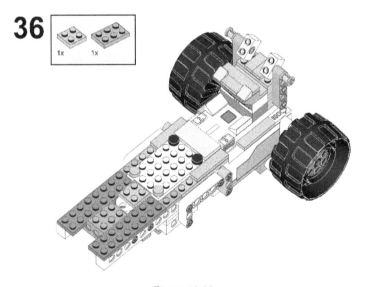

Figure 12.38

37. Take four 1x2 plates and connect them so that they're between the black and orange round flat tiles. Then, take two 1x4x2/3 plates with bows and connect them to the 2x2 plate, as shown in the following figure:

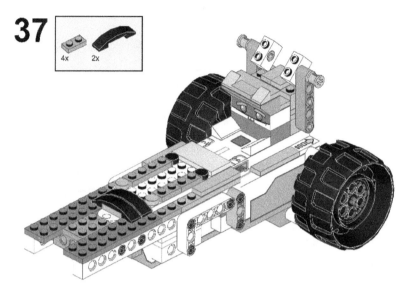

Figure 12.39

38. Now, take one 1x6 brick and one 1x6 brick with a bow and connect them to the plates, as shown in the following figure:

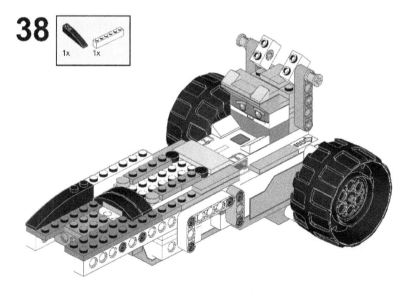

Figure 12.40

39. Then, take one 1x8 flat tile and connect it to both the 1x6 brick and the 1x6 brick with a bow:

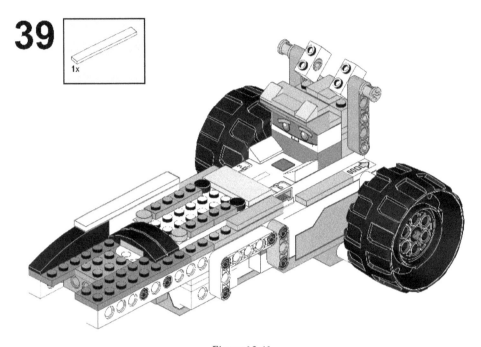

Figure 12.41

40. Now, let's make the connections to the external motor. To do this, take an external motor from the BOOST kit:

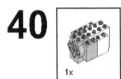

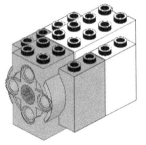

Figure 12.42

41. Then, take two connector bushes with axles and one 3M axle and connect them to the motor, as shown in the following figure:

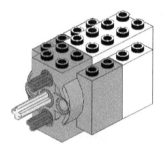

Figure 12.43

42. Take two 1x4 flat tiles and connect them to both sides of the motor, as shown here. Then, take one 3M beam with a fork and connect it to the 3M axle. Once you've done that, connect both connector bushes to the axle:

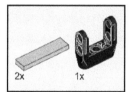

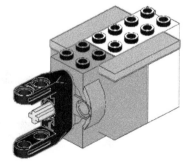

Figure 12.44

43. Then, take one Z12 conical wheel and connect it to the 3M axle. Once you've done that, take one 2x2 plate with a knob and one 2x4 flat tile and connect both to the external motor, as shown in the following figure:

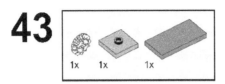

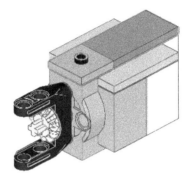

Figure 12.45

44. Take one 1x1 angular brick and connect it to the plate with a knob. Then, take one orange 1x1 round flat tile and one black 1x1 flat tile and connect them to the angular brick:

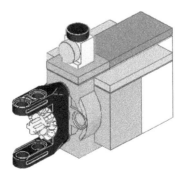

Figure 12.46

45. Now, connect this motor to all four 1x2 plates that are connected to the BOOST Hub, as shown in the following figure:

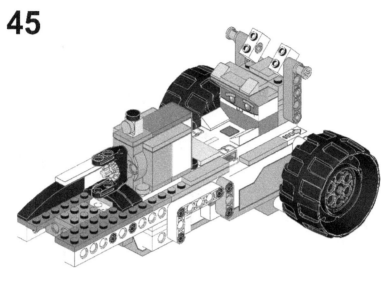

Figure 12.47

46. Now, take one bevel gear and place it on the beam with a fork so that it will mesh with the conical wheel, as shown in the following figure. Then, take one catch with a cross hole and one 8M stopper axle. Pass this axle through the beam with a fork, the catch with a cross hole, and the bevel gear, as shown here:

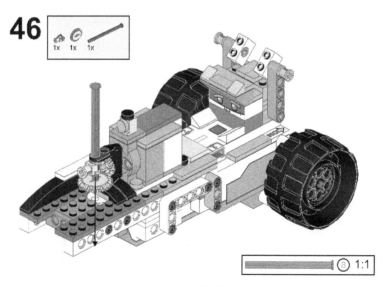

Figure 12.48

47. Now, take one 1x6 brick and one 1x6 brick with a bow and connect them to the plates, as shown in the following figure:

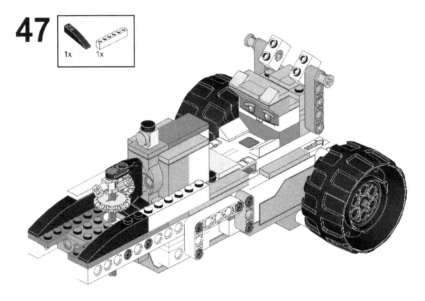

Figure 12.49

48. Next, take one 1x8 flat tile and connect it to both bricks. Then, take two 2x4 bricks and connect them so that they're between the bricks with bows, as shown in the following figure:

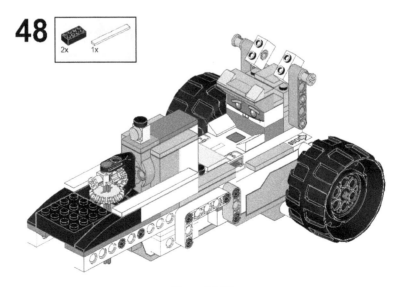

Figure 12.50

49. Now, take one 1x4 plate and one 3x4x2/3 plate with a bow and knobs and connect them to both 2x4 bricks, as shown in the following figure:

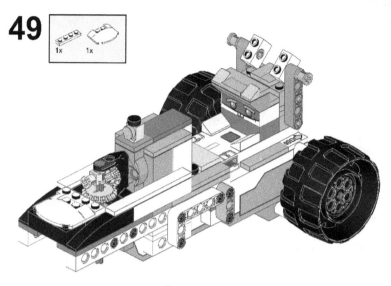

Figure 12.51

50. Then, take one round plate with a knob and connect it to both the 1x4 plate and the plate with a bow and knobs. Then, take three orange 1x1 round flat tiles and connect two of them to the 1x4 plate. Connect the remaining one to the round plate with one knob, as shown here:

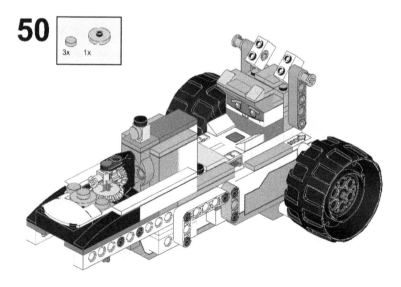

Figure 12.52

51. Take one 1x4 plate and one 1x2 plate and connect the 1x2 plate to the 1x4 plate:

Figure 12.53

52. Now, take two white 1x1x1 bricks with arches and connect them to both sides of the 1x2 plate:

Figure 12.54

53. Now, take two blue 1x1x1 bricks with arches and connect them to the 1x2 plate, as shown in the following figure:

Figure 12.55

54. Now, take one 2M axle and connect it to the catch with a cross hole, as shown here:

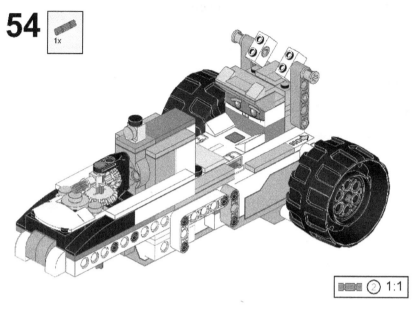

Figure 12.56

55. Now, let's make the front wheels. Take two 2x2 round bricks with holes and two 4M stopper axles. Pass both axles through the bricks:

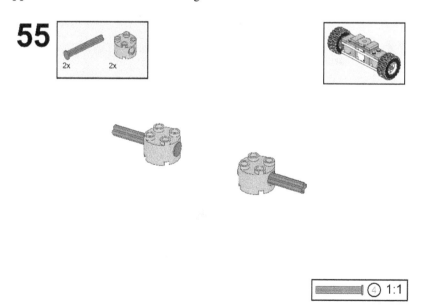

Figure 12.57

56. Take two ½ bushes and connect them to both the stopper axles. Then, take one 2x8 plate with a hole in its side and connect both bricks via the hole on both ends of the plate, as shown here:

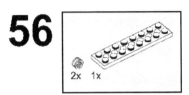

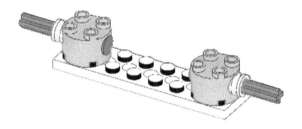

Figure 12.58

57. Take two 1x2 bricks and connect them to both bricks with holes. Then, take one brick with a cross and connect it to the center part of the plate:

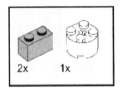

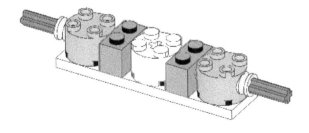

Figure 12.59

58. Again, take one 2x8 plate with a hole in its side and connect it to the upper part of the bricks:

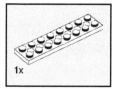

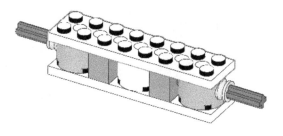

Figure 12.60

59. Now, take two 1x2 flat tiles and connect them to both ends of the 2x8 plate. Then, take two 2x2 plates and connect them to the same plate, beside the flat tiles, as shown here:

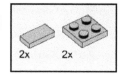

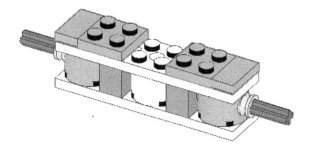

Figure 12.61

60. Take two rims with crosses and two tires. Connect the tires to the rims and then connect both completed components to both stopper axles, as shown in the following figure:

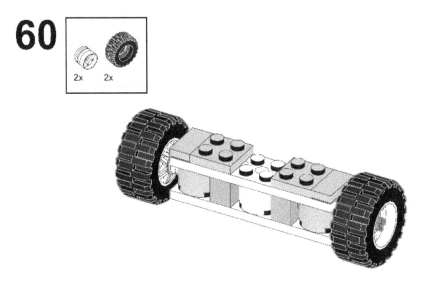

Figure 12.62

61. Now, take one 2x2 round plate and connect it to the center part of the 2x8 plate. Then, take one 2x4 plate with a hole through it and connect it to all three plates – that is, the two 2x2 plates and the one round plate, as shown in the following figure:

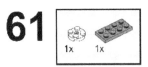

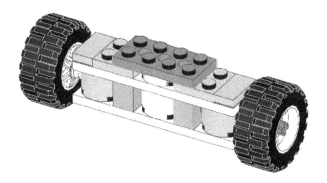

Figure 12.63

62. Take one 2x2 round flat tile with a hole through it and connect it to the center part of the 2x4 plate. Then, take two 1x2 flat tiles and connect them to both sides of the round flat tile:

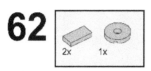

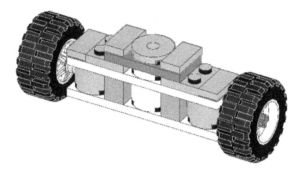

Figure 12.64

63. Now, the front wheels are ready to be connected to the main model. To do this, pass the 8M stopper axle through the center hole of the front part that we have just created:

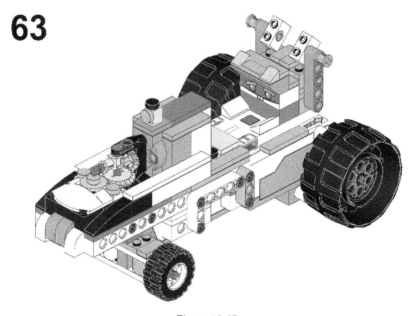

Figure 12.65

64. Now, let's decorate our car. To do so, take one 2x6 plate, like so:

Figure 12.66

65. Then, take one 3x8 right angle plate and one 3x8 left angle plate and connect them to the 2x6 plate, as shown in the following figure:

Figure 12.67

66. Take two 1x6 bricks with bows and two 2x6 bricks with bows and connect them to the left- and right-angle plates:

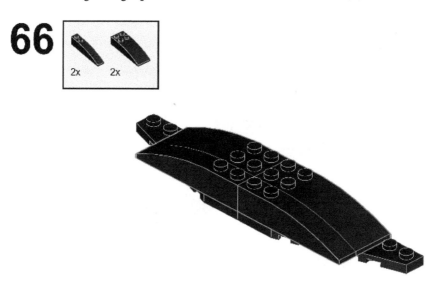

Figure 12.68

67. Next, take one 3x8x2 left shell with a bow and angle and one 3x8x2 right shell with a bow and angle. Connect them as shown in the following figure:

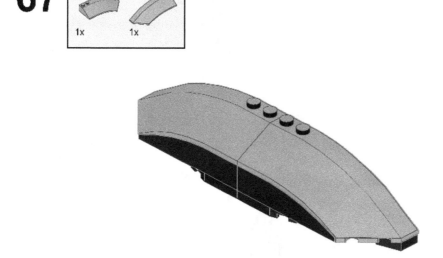

Figure 12.69

68. Take one 1x4 flat tile and connect it to both the left and right shells:

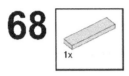

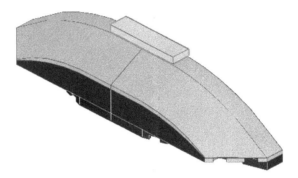

Figure 12.70

69. Now, connect that structure to both white 1x2 bricks, as shown in the following figure:

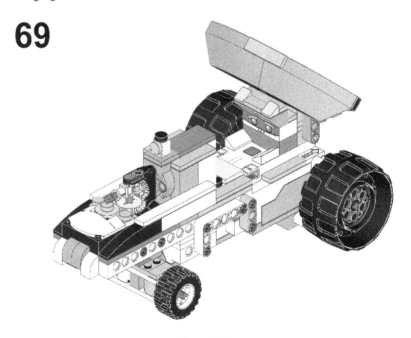

Figure 12.71

Great! Your racing car is ready to use!

Now, let's code it and make it run. Check that your car looks as follows:

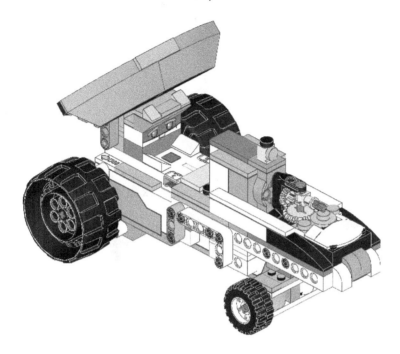

Figure 12.72

Connect the external motor to port C.

Now, let's complete some programming tasks.

Let's code the robot to run on different racetracks

First, find out the number of rotations of motor C that are needed for left steering and right steering:

- Number of rotations for left steer = _____ rotations.
- Number of rotations for right steer = _____ rotations.

Now, let's do some interesting activities.

Activity #1

Try to use some items in your home and form a race track that looks as follows. Get to the end of this race track using your coding skills:

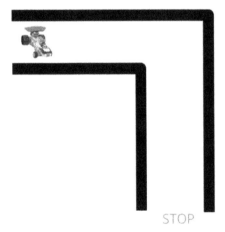

Figure 12.73 – Race track setup 1

Let's move on to the next activity.

Activity #2

For this activity, simply swap the start and stop positions shown in *Activity #1* and run the robot on same race track again:

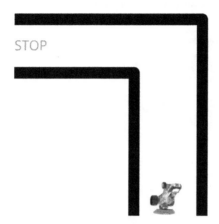

Figure 12.74 – Race track setup 2

Let's try another activity with the race track.

Activity # 3

Now that you know how to turn left and right, take some extra items from your home, and form the race track shown in the following figure. Try to finish this in the shortest amount of time possible:

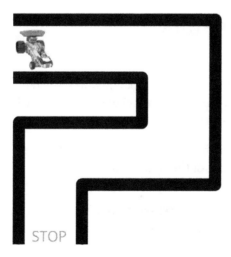

Figure 12.75 – Race track setup 3

How quickly could you finish this route? _____ seconds.

The sample code for the activity is given here:

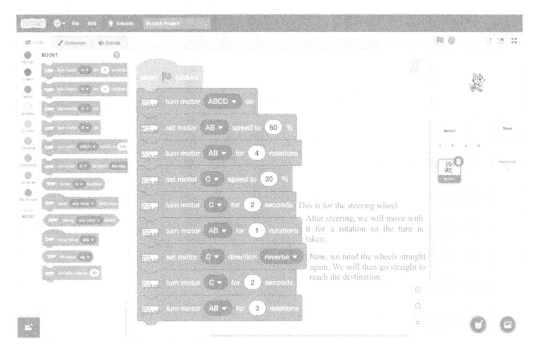

Figure 12.76 - Sample code

Now, let's try to solve a challenge.

Time for a challenge

Challenge #1

Attach a color sensor to the front of the robot and find some red and green objects that your sensor can sense. Now, create the track shown in the following image and get to the end of the racetrack while ensuring the following conditions are satisfied:

- The car must only start the journey when it senses the green object. You must place this green object in front of the color sensor when you wish to start the journey on the racetrack.

- The car must stop whenever it senses the red object. Try to show this object at the stop point of the racetrack.

Hint: Position your color sensor horizontally so that it can sense the objects that you have provided.

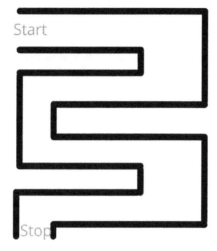

Figure 12.77 – Final racetrack

You can try many such different routes and try to finish them as quickly as possible. This will help you enhance your coding skills as well.

Summary

In this chapter, you built a race car with a steering wheel and navigated various race tracks using the coding blocks that you learned about in the previous chapters. You also enhanced your building skills by attaching the color sensor to the front of the robot on your own to solve this chapter's challenge. You can now apply advanced coding concepts to build and code completely autonomous cars, as well as other robotics applications with sensors mounted on them. With this, we have come to the end of our exciting journey of learning how to build and code using the LEGO BOOST kit.

In the next chapter, you will be solving a challenge without any guidance provided on how to build and code the robot. This will help you figure out what level your learning is at regarding this kit and will help you learn more if you're stuck on something. It is highly recommended that you go through this and the previous chapters before you move on to the next chapter. All the best!

Further reading

Did you know that it takes less than 3 seconds for the technicians at the pit stop in F1 races to change the wheels of any F1 race car? To find out more about this and many more exciting facts about F1 racing, go to

```
https://www.thrillist.com/cars/facts-about-race-cars-
surprising-details-about-the-world-s-fastest-vehicles.
```

13
Final Challenge

Did you know that by applying the LEGO construction and coding skills, you can participate in various international robotics competitions such as World Robot Olympiad, FIRST LEGO League, Robotex, and others? These are great platforms to express your creativity, innovation, logical thinking, and problem-solving skills, and they also help you get great experience working as a team.

In this chapter, you will be solving a similar challenge using your LEGO BOOST kit. Use the following setup for your robot and then read the rules carefully before you start building and coding your robot:

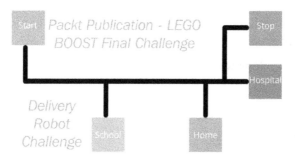

Figure 13.1 – Final challenge setup

In this chapter, we will cover the following topics:

- Building the robot
- Let's code

You must build a robot that can carry three 2x4 LEGO bricks – one red and two black. The red brick represents food, while the black brick represents medical essentials. The robot needs to do the following:

1. Start from the starting point. The robot's size must not exceed 35 cm x 35 cm x 35 cm. Load all three LEGO bricks/food, as well as medical essentials, before the robot starts its journey.

2. Deliver the medical essentials to a school and a hospital.

3. Deliver the food to the appropriate home.

4. Stop the robot when it reaches the stop zone.

Following are the rules of the challenge:

1. The maximum time to finish this challenge is 180 seconds.

2. You get 20 points for dropping the correct thing at the correct destination.

3. You get 30 points for parking correctly.

4. You get 10 points for stopping the robot completely in the stop zone. This means that you can score a maximum of 100 points in this challenge.

Now, let's get started!

Technical requirements

In this chapter, you will need the following:

- A LEGO BOOST kit with six AAA batteries, fully charged
- A laptop/desktop with the Scratch 3.0 programming language installed and an active internet connection
- A diary/notebook, along with a pencil and eraser
- Black tape, 1 inch wide

Building the robot

Answer the following questions once you have understood the challenge properly:

1. Will you need a two-wheel drive or a one-wheel drive robot? _____

2. Will you build a back-wheel drive or a front-wheel drive robot? _____

3. Which type of mechanism will you need to carry three LEGO bricks with the robot? A conveyer belt, a grabber, or a dispenser? _____

4. Do you need to follow the line to finish this task accurately? Yes or no? _____

Now, let's start building the robot that will crack this challenge. Remember that you will never get the final solution on the first try. Keep trying until you get a robot with a functional dropping mechanism.

Let's code

Always break your bigger problem into multiple small problems and solve them one after another. This will help you to solve tasks quickly and easily.

Activity #1

Write some code so that the robot can reach the desired school from the starting point. Drop the medical essentials at this place.

Activity #2

Now, write some code so that the robot will travel from the school to the desired home and drop the food there.

Activity #3

Now, write some code so that the robot will travel from the desired home to the hospital and drop the medical essentials there.

Activity #4

Reach the stop position and stop the robot.

Activity #5

Now, club all these small programs together and try to crack the complete challenge. Your code must be accurate enough that it does not need any human intervention from start to stop. Try to solve this challenge slowly at first. Eventually, increase the speed of the robot and try to crack this challenge in as little time as possible.

Summary

In this chapter, you built a robot to crack the challenge outlined. You learned how to use different mechanisms effectively to solve the tasks at hand. You also applied your coding skills to ensure that the robot finishes all the tasks as needed in the given time slot. Try to create such challenges on your own and solve them to sharpen your construction and coding skills. You can also consider using the LEGO BOOST mobile app so that you can use the sensor in distance sensing mode too. In the next chapter, we will learn about the grabbing robot.

Further reading

To learn more about various robotics competitions such as World Robot Olympiad and FIRST LEGO League, look at the following links. See whether you can team up with your friends/form a team at your school and participate in these competitions:

- WRO: https://wro-association.org/home
- FIRST LEGO League: https://www.firstinspires.org/robotics/fll

Bonus Chapters

14
The Grabbing Robot

In the previous chapter, you built a compact forklift and programmed it to move boxes from one place to another. Did you notice that the forklift was unable to carry unevenly shaped objects or even circular objects properly from one place to another? Well, forklifts can easily lift square and rectangular boxes, but to pick up small objects, circular objects, or anything else that's an obscure shape, we need to implement a grabber with a hand-like grip that can easily grab and drop things.

In this chapter, you will be building a similar grabbing mechanism and completing various tasks. Did you know that your hand is an example of a simple machine lever? Your wrist, palm, and fingers act as a load-carrying agent, your elbow is the fulcrum, and they work together. You will be implementing the same concept while building this grabber and replicating the movement of your hand. The following figure shows a robot with a grabbing arm like the one we will build in this chapter:

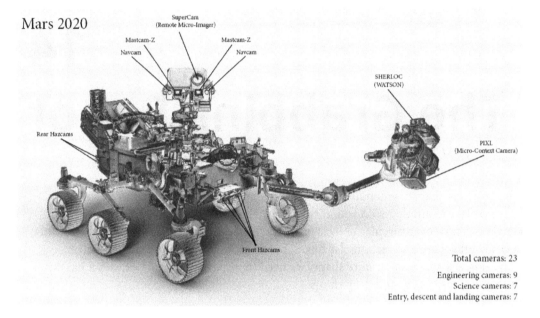

Figure 14.1 – A grabbing arm connected to a robot

In this chapter, we will cover the following topics:

- Building the grabbing robot
- Let's code the robot to grab, displace, and drop a bottle
- Time for a challenge

Technical requirements

In this chapter, you will need the following:

- A LEGO BOOST kit with 6 AAA batteries, fully charged
- A laptop/desktop with the Scratch 3.0 programming language installed and an active internet connection

- A diary/notebook, along with a pencil and eraser
- Basic items from your home, such as a water bottle and lunch box

Building the grabbing robot

Refer the following link for the detailed build instructions of the grabbing robot:

```
https://github.com/PacktPublishing/Build-and-Code-Creative-
Robots-with-LEGO-BOOST/blob/main/Bonus%20Chapter/Chapter14/
B17095_BuildInstructions_14_Final_NM.pdf
```

The grabbing robot will look like this:

Figure 14.2

Once you have finished building this robot, connect the external motor to port C. Let's get started with the various programming tasks we must complete to create this robot.

Let's code the robot to grab, displace, and drop a bottle

Take a half-liter water bottle and fill it with water. Now, implement the following setup:

Figure 14.3 – Route setup

Let's start the activity!

Activity #1

Program your robot to grab the bottle from its position and drop it at the given drop location. Wait for 2 seconds after grabbing the bottle and display **Bottle grabbed successfully**. After dropping the bottle, display **Bottle dropped successfully**. You can play with the brick's light color and the sound blocks when various actions are performed. Try to solve this problem in the fastest time possible.

The sample code for the activity is given here:

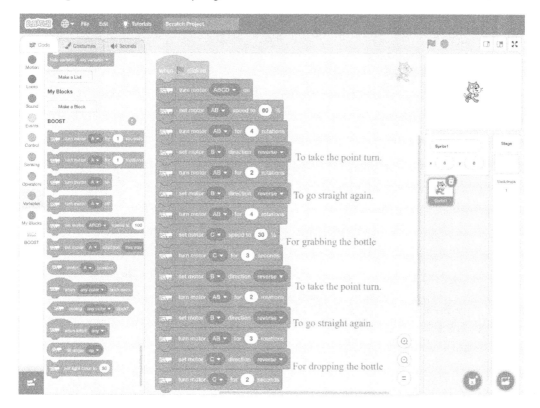

Figure 14.4 - Sample code

Now, let's try a new challenge

Time for a challenge

Challenge #1

Complete the following task in the shortest possible time. The robot needs to do the following:

1. Grab a bottle and drop it at the respective location.
2. Grab a pen/pencil holder and drop it at the respective location.

Ensure that you have the following setup for this exercise:

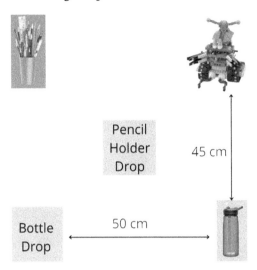

Pencil Holder Drop

45 cm

Bottle Drop

50 cm

Figure 14.5 – Route for the challenge

This challenge involved coding. Now, let's look at the next challenge, which will involve construction.

Challenge #2

Create a raised platform using either the books in your home or extra LEGO bricks. Now, create the same route as shown in the preceding figure and try to place both objects on the raised platform so that the objects do not fall. Use some creativity when creating these raised platforms.

Summary

In this chapter, you were introduced to a pick-and-place grabber using a single motor. Such grabbers can be used in various places to pick up all kinds of objects. You also applied your coding skills to solve the challenges in this chapter and finish them as quickly as possible.

In the next chapter, you will be building an obstacle avoidance robot using color sensor that comes with your BOOST kit. You will learn to program the robot with sensors on it.

Further reading

- You can learn more about grabbing robots at `https://www.nature.com/articles/d41586-018-05093-1`.

- Did you know that *Asimo*, the first humanoid robot, can hold objects carefully? It knows how much pressure to apply to different kinds of objects. For example, if it holds a paper object too firmly, it knows that it will break. So, it will only hold the item with the required pressure! You can learn more about the functionality of *Asimo* at `https://youtu.be/6lmh32xLvJE`.

15
Obstacle Avoidance Robot

Until now, you have been writing simple programs for your robot using basic instructions. What if you want to make a real robot that can perform tasks autonomously? What makes the robot take decisions and then act upon them? It is the sensors on the robot that help it through this entire process. Different types of sensors help robots during different tasks. These sensors also make them act intelligently in given situations. For example, an ultrasonic sensor, along with a camera, offers vision to a robot. A touch sensor offers touch-type sensations to a robot, and so on. In this chapter, you will learn about the function and application of a color sensor to build an obstacle avoidance robot. A sensor that helps the robot to identify colors correctly and take respective decisions is called a color sensor. Color sensors are widely used for line-following robots, color detection and sorting application-based robots in industries, self-driving cars, and more. In this project, you will build a robot with a color sensor on it, then learn to make your robot perform different actions based on the color that it detects.

In this chapter, we will cover the following topics:

- Building an obstacle avoidance robot
- Let's code the robot to avoid obstacles
- Time for a challenge

Technical requirements

In this chapter, you will need the following:

- A LEGO BOOST kit with six AAA batteries, fully charged

- A laptop/desktop with Scratch 3.0 programming and an active internet connection

- A diary/notebook with a pencil and eraser

- Red and green paper

Building the obstacle avoidance robot

Refer the following link for the detailed build instructions of the obstacle avoidance robot:

```
https://github.com/PacktPublishing/Build-and-Code-Creative-
Robots-with-LEGO-BOOST/blob/main/Bonus%20Chapter/Chapter15/
B17095_BuildInstructions_15_Final_NM.pdf
```

The robot will look as follows:

Figure 15.1 – Obstacle avoidance robot

Let's now move on to the coding section and perform the activities listed.

Let's code the robot to avoid obstacles

Your LEGO BOOST kit's color sensor can detect six different colors: red, blue, green, yellow, black, and white. It can detect color from a maximum distance of 3 cm. Now, create the following setup before we start with the activities:

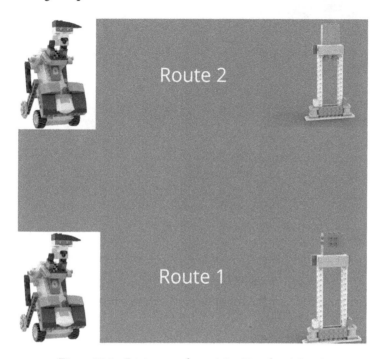

Figure 15.2 – Route setup for activity #1 and activity #2

Let's get started.

Activity #1

Go forward until the robot senses a red pole and change the brick light to red. Let's first write an algorithm to reflect this:

1. Move forward at 20% speed.

2. Wait until red is sensed.

3. Stop motor AB when red is sensed.

4. Change the brick light color to red.

The sample code for this activity is as shown in the following figure:

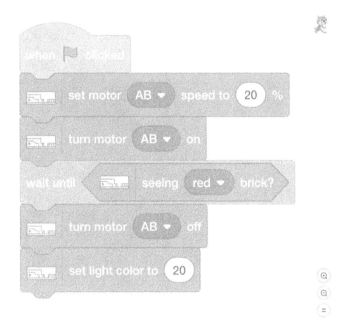

Figure 15.3 – Sample code

Let's now write a similar code for the robot to execute when it senses the blue brick.

Activity #2

Program your robot to turn left whenever it senses blue. Write an algorithm on your own and then execute it using the Scratch 3.0 programming language.

Let's now solve a challenge to test your knowledge and understanding.

Time for a challenge

Challenge #1

Let's make a color detection-based pathfinder robot. Program your robot to do the following:

1. Move forward whenever no color is detected.
2. If red is detected, turn left. Change the hub light color respectively.
3. If blue is detected, turn right. Change hub light color respectively.
4. Stop when the robot reaches the endpoint. Change the hub light color to green.

Create the following setup in your home:

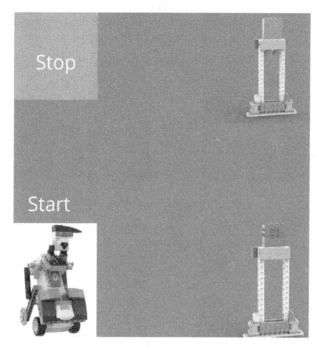

Figure 15.4 – Challenge setup

You can be creative and add sound and display blocks as needed to make solving this challenge more interesting.

Summary

In this chapter, you were introduced to sensor-based programming, and you learned about the functioning of the LEGO BOOST color sensor. It senses six distinct colors. You performed coding exercises to make your robot perform various actions based on the color that it detects. One important thing learned in this chapter is that we never use rotations or second-based programming when we want our actions to be decided based on inputs received from the sensor. In the next chapter, you will be building a line-following robot that can follow the edges of a black line. This robot will look like a humanoid and its programming will be logical. You will learn about the importance of line-following robots in warehouses. You will also solve tasks involving line following.

Further reading

If you use LEGO BOOST's Android mobile app or iPad app, you can use this color sensor as a distance sensor. This will make these activities a lot simpler, as it can detect obstacles at a distance and take actions based on the programming logic that you have set. You can download this app at the following URL:

`https://www.lego.com/en-us/kids/games/boost/lego-boost-03c8a71 ab07a428fba9fa3b460b387b2`

To learn more about this app, visit the following URL: `https://www.lego.com/ en-us/service/help/products/electronics-robotics/boost/guide- to-lego-boost-programming-blocks-408100000009897`.

You can use the same pillars as obstacles and program your robot to stop when an obstacle is detected, then take an action.

16
The BOOST Humanoid

Up to this point, you have built various types of robots using this book. Did you observe one thing that they have in common? None of these robots have their own eyes through which they can see and follow a specific path! Usually, robots tend to follow the line and accomplish the iterative tasks on a specific given route. Ideally, these robots are used in warehouses as well as other indoor industrial applications. You will be building a similar line-following robot using the color sensor that you have in the BOOST kit. As the name suggests, a robot will follow the black line and reach out to the required places to accomplish the tasks:

Figure 16.1 – Line-following robots at a warehouse

In this chapter, you will do the following:

- Build a line-following robot
- Let's code a robot to follow the line
- Time for a challenge

In this robot, you will be replicating a human-like structure with legs, hands, and a head. Give this robot some cool name of your choosing.

Technical requirements

In this chapter, you will need the following:

- LEGO BOOST kit with six AAA batteries, fully charged
- A laptop/desktop with Scratch 3.0 programming and an active internet
- One-inch-thick black tape
- A table top with a white background

Building a line-following robot

Refer the following link for the detailed build instructions of the line-follow robot:

```
https://github.com/PacktPublishing/Build-and-Code-Creative-
Robots-with-LEGO-BOOST/blob/main/Bonus%20Chapter/Chapter16/
B17095_BuildInstructions_16_Final_NM.pdf
```

Your line following robot will look as shown in the following figure:

Figure 16.2

Let's now code this robot to follow the line.

Let's code the robot to follow the line

Let's now start coding this line-following robot. There are various line-following techniques, such as **two-state**, **five-state**, **proportional**, and **proportional integral derivative**. In this chapter, you will be learning the easiest yet most accurate technique – the two-state line follow. You will be using the **if-else** programming block from the **Control** pallet. For line-following robots, it is necessary to have the right environment, including the following:

- A plain white background
- A properly etched black line
- Controlled lighting conditions

Remember one thing. The robot always follows the edge of the line and not the center. Wondering why? This is because, if the robot follows the center of the line, it will miss the turns and leave the line, but if it is following the edge, it will never ever miss the turns/curves. The algorithm for two-state line following is simple – *if the robot senses black, turn toward the white, otherwise turn toward the black*. This will ensure that the robot is always near the edge while following the black line. Do the following setup before you start with the activities. Use the one-inch-thick tape on the white background to do this setup:

Figure 16.3 – Field setup

Let's now perform an interesting activity.

Activity #1

Write a code to follow the line for 5 seconds at 20% motor power:

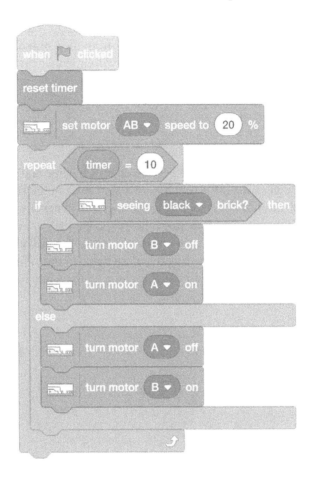

Figure 16.4 – Sample code

You will be able to follow the right edge of the line using this code. Simply swap the motor ports in the **if-else** condition and the robot will follow the left edge of the line. If you increase the speed of this robot, it will wobble more and cover more distance in the same time. More wobbling is not good for line following, hence we have retained a slow speed for the robot.

Time for a challenge

Challenge #1

Attach the external motor to this robot on the top and build an arm that can hit the tennis ball. Now, perform the following setup:

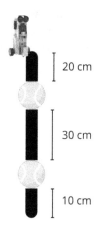

Figure 16.5 – Route setup

Solve the following tasks:

1. Follow the line until the first ball. Now, hit the ball and clear the robot's path to reach the next stop. Change the BOOST Hub color to red.

2. Now, follow the line to the second ball and hit that ball as well. Change the BOOST Hub color to blue.

3. Reach the stopping place and play a sound of your choice.

The robot must finish this challenge accurately each time. While following the line, you must focus on accuracy instead of speed, as the most important element in such cases is the accuracy of the tasks solved, followed by time. You can play around with the speed of this robot and see how quickly you can complete this challenge with the utmost accuracy.

Summary

In this chapter, you learned about the practical application of the line-following technique and its use in industry. You wrote your own two-state line-following program and executed the same to solve various tasks. You also applied your construction skills in the challenge section to solve the challenge given. You will be able to build such humanoid robots on your own using LEGO elements on any other platform, such as *EV3* and *WeDo*. In the next chapter, you will be building a robot with wheels and a color sensor on it that will be landing on the moon's surface for exploration purposes. You will be performing logical activities to identify the types of minerals present on the surface of the moon and grab some of these minerals.

Further reading

Amazon is said to be leading the race when it comes to automation and robotics at its gigantic warehouses across the globe. Read the following article about their execution of line-following robots: `https://www.wired.com/story/amazon-warehouse-robots/`.

Once you are familiar with programming, you can learn about advanced line-following techniques such as five-state and proportional derivative integrals using advanced Mindstorms sets or any other open-source platforms.

17
The Moon Rover

We all live on planet Earth! But did you know that there are millions of planets in space? Although we have fantasized about having life like humans on other planets, we have never seen any until now. Scientists across the globe are trying hard to find traces of life and other resources on different planets every day! We have already been successful in sending robots to Mars and our very own moon to carry out various research projects, such as finding oxygen, water, and minerals. These days, scientists are trying to establish the fact that humans can survive on other planets as well! Elon Musk is already planning to have a habitat for humans on Mars by 2025 and send humans to live there. Recently, India sent a rover to the moon that crash-landed just 4 seconds before the actual landing. If the mission had succeeded, it would have been the cheapest space mission in history.

In this chapter, you will build a similar rover and explore some interesting areas on the moon:

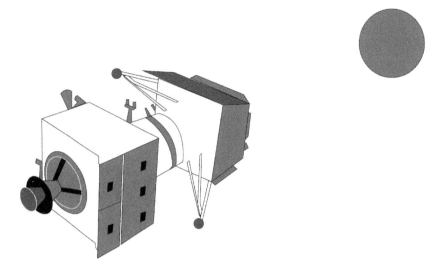

Figure 17.1 – Vikram lander carrying the Pragyan rover

In this chapter, we will cover the following topics:

- Building the moon rover
- Let's code the robot to perform different tasks
- Time for a challenge

Since the terrain in such places is uneven, having track wheel drive will help us keep the robot moving in such conditions, without falling into small pits or bumps. Hence, we will be using a track wheel drive for our robot.

Technical requirements

In this chapter, you will need the following:

- A LEGO BOOST kit with 6 AAA batteries, fully charged
- A laptop/desktop with the Scratch 3.0 programming language installed and an active internet connection
- A diary/notebook, along with a pencil and eraser
- Blue, green, and yellow sticky notes

Building the moon rover

Refer the following link for the detailed build instructions of the moon rover:

```
https://github.com/PacktPublishing/Build-and-Code-Creative-
Robots-with-LEGO-BOOST/blob/main/Bonus%20Chapter/Chapter17/
B17095_BuildInstructions_17_Final_NM.pdf
```

The moon rover will look like this:

Figure 17.2

In the next section, we will complete some interesting activities with this moon rover.

Let's code the robot to perform different tasks

This robot is on a mission to identify various minerals on the surface of the moon. First, use your sticky notes and create the following setup:

Figure 17.3 – Route setup

Here, blue represents iron, green represents copper, and yellow represents gypsum.

Activity #1

Write a program so that your rover does the following:

1.   Stops on any colored line (blue, green, or yellow) on its route.

2.   If blue is detected, wait for 2 seconds and display **Iron found!**.

3.   If green is detected, wait for 2 seconds and display **Copper found!**.

4.   If yellow is detected, wait for 2 seconds and display **Gypsum found!**.

The sample code for the activity is given here:

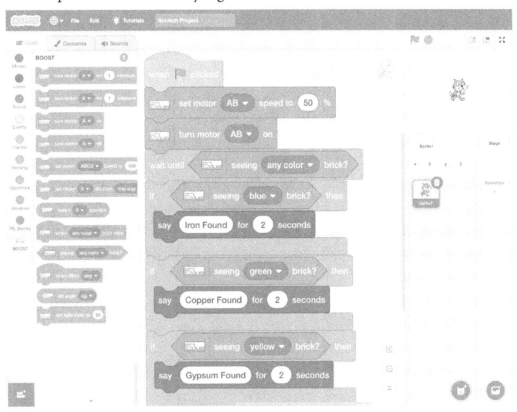

Figure 17.4 - Sample code

You can use sound blocks at the respective locations to make this project more interactive. The program should work for all iterations of the colored sticky notes. Any sequence should work.

Time for a challenge

In this section, you will build some game elements from LEGO bricks on your own. Your robot already has a grabber hand attached to it. You must build three elements that represent copper, iron, and gypsum, respectively. The grabber should be able to pick up these three elements. Now, set up the route shown in the following diagram. These elements should be called iron ore, copper ore, and gypsum ore:

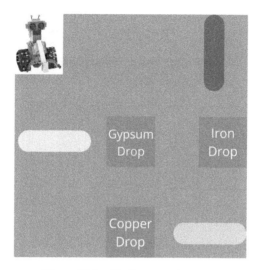

Figure 17.5 – Challenge route set up

Now, let's get started!

Challenge #1

After completing this setup, write some code that will make the robot do the following:

1. Pick up the ore from their respective locations and drop them at their drop locations.
2. Display the pickup and drop on the screen.
3. Use sound blocks appropriately.
4. If possible, play with the BOOST Hub brick light as well.

You can try to add more features to this robot and try to complete more interesting tasks on the surface of the moon with this rover.

Summary

In this chapter, you learned how space missions work. Usually, they have extremely complex control systems, construction, and coding as they must send signals from millions of miles away to the Earth. Besides this, they must combat extremely hot and cold temperatures throughout their journey. You built a rover with track wheel drive and completed missions to identify minerals on the moon.

Further reading

You can learn more about India's mission moon (*Chandrayaan 2*) at `https://www.isro.gov.in/chandrayaan2-home-0`.

The failure of this mission has not deterred the intentions of scientists at ISRO, and they are already preparing for *Chandrayaan 3*, with its proposed launch being in November 2022.

Packt.com

Subscribe to our online digital library for full access to over 7,000 books and videos, as well as industry leading tools to help you plan your personal development and advance your career. For more information, please visit our website.

Why subscribe?

- Spend less time learning and more time coding with practical eBooks and Videos from over 4,000 industry professionals

- Improve your learning with Skill Plans built especially for you

- Get a free eBook or video every month

- Fully searchable for easy access to vital information

- Copy and paste, print, and bookmark content

Did you know that Packt offers eBook versions of every book published, with PDF and ePub files available? You can upgrade to the eBook version at packt.com and as a print book customer, you are entitled to a discount on the eBook copy. Get in touch with us at customercare@packtpub.com for more details.

At www.packt.com, you can also read a collection of free technical articles, sign up for a range of free newsletters, and receive exclusive discounts and offers on Packt books and eBooks.

Other Books You May Enjoy

If you enjoyed this book, you may be interested in these other books by Packt:

Learn Robotics Programming - Second Edition

Danny Staple

ISBN: 978-1-83921-880-4

- Leverage the features of the Raspberry Pi OS
- Discover how to configure a Raspberry Pi to build an AI-enabled robot
- Interface motors and sensors with a Raspberry Pi
- Code your robot to develop engaging and intelligent robot behavior
- Explore AI behavior such as speech recognition and visual processing
- Find out how you can control AI robots with a mobile phone over Wi-Fi
- Understand how to choose the right parts and assemble your robot

Smart Robotics with LEGO MINDSTORMS Robot Inventor

Aaron Maurer

ISBN: 978-1-80056-840-2

- Discover how the Robot Inventor kit works, and explore its parts and the elements inside them
- Delve into the block coding language used to build robots
- Find out how to create interactive robots with the help of sensors
- Understand the importance of real-world robots in today's landscape
- Recognize different ways to build new ideas based on existing solutions
- Design basic to advanced level robots using the Robot Inventor kit

Packt is searching for authors like you

If you're interested in becoming an author for Packt, please visit `authors.packtpub.com` and apply today. We have worked with thousands of developers and tech professionals, just like you, to help them share their insight with the global tech community. You can make a general application, apply for a specific hot topic that we are recruiting an author for, or submit your own idea.

Share Your Thoughts

Now you've finished *Build and Code Creative Robots with LEGO BOOST*, we'd love to hear your thoughts! Scan the QR code below to go straight to the Amazon review page for this book and share your feedback or leave a review on the site that you purchased it from.

https://packt.link/r/1801075573

Your review is important to us and the tech community and will help us make sure we're delivering excellent quality content.

Index

www.ingramcontent.com/pod-product-compliance
Lightning Source LLC
Chambersburg PA
CBHW081501050326

40690CB00015B/2881